D0563979

MODERN SCIENCE AND THE PARANORMAL

HAUNTED:
GHOSTS AND THE PARANORMAL

MODERN SCIENCE AND THE PARANORMAL

MARIE D. JONES

ROSEN
PUBLISHING®

New York

This edition published in 2009 by:

The Rosen Publishing Group, Inc.
29 East 21st Street
New York, NY 10010

Cover design by Nelson Sá

Library of Congress Cataloging-in-Publication Data

Jones, Marie D., 1961–
[PSIence]
Modern science and the paranormal / Marie D. Jones.
    p. cm.—(Haunted)
Includes bibliographical references and index.
ISBN-13: 978-1-4358-5179-5 (library binding)
1. Parapsychology and science. I. Title.
BF1045.S33J66 2009
130—dc22

2008035686

Manufactured in the United States of America

First published as *PSIence: How New Discoveries in Quantum Physics and New Science May Explain the Existence of Paranormal Phenomena* by New Page Books / Career Press, copyright © 2007 by Marie D. Jones

Photo credits: cover (right), pp. 2, 5 © istockphoto.com/Michael Knight; cover (left) © istockphoto.com.

For Max, my Theory of Everything.

DEDICATION

7

ACKNOWLEDGMENTS

Nobody writes a book alone. Well that's not really true. I wrote this book alone. I don't recall anyone else stepping in to write a chapter or two. But like a good pair of panty hose, I had a lot of support. My team consists of some great cheerleaders who have been behind me all the way.

First and foremost my mom, Milly—best friend, confidante, and ultimate babysitter supreme, and my dad, John—scientist, pal, editor extaodinaire, and fellow booklover, for their love, support and belief, in all my dreams, even the strange ones. To my husband, Ron, my best friend and the guy who makes me laugh and supplies me with chocolate when I am stressed out, which is daily. To my extended family, friends, and colleagues, who give me wings. Thanks to Ron Jones and Aaron "Owen" Hayhurst (*www.owensedge.com*) for the awesome graphics! And thanks to Winnie for my Web site—thanks!

To everyone involved in the book design, editing, production, and promotion, you guys and gals are top-notch.

To the two Lisas—Lisa Hagan of Paraview, Inc., my amazing agent, friend, and co-conspirator who knows exactly what I am all about and won't let me settle for less. And Lisa Collazo of WriteWhatYouKnow.com, the best creative coach out there, and the one who keeps me sane and on track! And to Andrea, my column of support!

To everyone who was so incredibly generous in offering help, ideas, research, and support, especially Nick Redfern, Pavel Mikoloski, Jim Marrs, Joshua P. Warren, Stephen Wagner, Hal Puthoff, and others who directed me right where I needed to be to get this book done.

To Sharon Shulz-Elsing of Curledup.com, for feeding my massive addiction to books. Some women crave furs, money, and jewelry. I prefer books. (Although I will take the money, too.)

To writers, researchers, and explorers everywhere who understand what I mean when I say there is nothing more thrilling than an unsolved mystery, with the exception of actually solving it.

To Helen—onward and upward!

And to the one who makes it all worthwhile, no matter what happens. Max. No scientist could ever explain, duplicate, or dissect the sheer force of love I feel when you take my hand and call me Mommy. Just please stop drawing on the new wood floors and don't kick the dog!

CONTENTS

In my travels, and in speaking to so many people around the world in my role as the grassroots marketing coordinator for the hit film *What the BLEEP Do We Know!?*, I discovered how grateful people were to be given permission to come out of the closet, if you will, in respect to their own experiences with "paranormal" phenomena. Because of *What the BLEEP*, people who had never before felt comfortable talking about their psychic intuition, their unusual encounter with a ghost or an astral being, or those who had seen a UFO, were now discussing their experiences freely. In so doing, they were counting themselves in on what has become, in essence, a cultural phenomenon not only in the United States, but around the world.

Much of the core audience for the film was comprised of those who had read a number of the books Marie refers to in the following chapters. The information is not new. What was different about *What the BLEEP* was the way the filmmakers put all the elements together. Here was the added power of film as an art form, combined with a story line in which scientific principles were incorporated into someone's life (Marlee Matlin, who played Amanda). The core audience members instinctively made the connection between their own lives and Amanda's—their own unusual experiences with the new science. Additionally liberating for people was the fact that, what they thought had been their isolated experience and understanding, was now a communal agreement, and in a big way. With a mainstream presentation of Western scientific under-pinnings for mystical experience, many who had felt relegated to the fringes now felt validated and a part of the larger whole.

Marie has been as much of a "fringe freak" as I have been. Similar to her, I was one of those "weird" children who consumed an enormous number of books on the subjects of PSI phenomenon, poltergeists, floating saints, astral projection, out-of-body experiences, UFOs, the phenomenon of spontaneous combustion, stigmata, time travel, and so many other phenomenal threads in a tapestry of mystery that con-founded our elders. We were on a quest to know—a quest that, at times, seemed insatiable; a quest that resulted in ridicule if we ventured to speak our minds about our out-of-body experiences, or our UFO sightings, or our encounters with "spirits."

Similar to Marie, I have been a student of esoterica, and yet an ardent student of the emerging new science of quantum physics and the expanding studies in consciousness. We both have an insatiable desire to know what the greatest minds of the world are discovering and proving in the realm of hard science.

To those who thought, similar to Marie and myself, that they were all alone in their quests to unlock life's mysteries, think again. Find joy in the discovery that there are others with similar journeys and similar hesitations in speaking the truth of their experiences. Relish the great minds and the excellent company you are now keeping!

In the year it was released, the little quantum fable by William Arntz, Betsy Chasse, and Mark Vicente, became one of the most successful documentaries in United States film history. I was told early on that there was "no audience" for What the BLEEP. But we were the audience, and we knew there were a lot more of us out there. This propelled me and the What the BLEEP marketing team to find the pockets of non-mainstream individuals and groups who would become the core audience for a film that is now a classic exploration into the union of science and spirit.

Marie was the first person I knew of to make the connection between the paranormal and hard science through quantum physics. She did it so beautifully, and so boldly that it leaves me with a true sense of admiration. This will be a landmark book for the disenfranchised who are, and always have been, sensitive to, and representative of, the quest for truth.

An expert layperson in communicating what can be some daunting new science, in *Modern Science and the Paranormal*, Marie has brilliantly condensed the essentials of quantum physics and made the subject matter not only understandable, but personally applicable. Her book is an exploration into the connections between what so many of us may have experienced and the science that is opening doors to a greater understanding and healing of the human dilemma. Hers is a welcome insight from what may soon no longer be the "borderlands" of psychic phenomenon, but rather a whole new realm of scientific enquiry that is satisfyingly mainstream.

—Pavel Mikoloski
What the BLEEP Do We Know!?
Grassroots Marketing Manager
Marketing Manager, Living the Field, London

There's an old saying in Hollywood: "Nobody knows anything." Credited to screenwriter William Goldman, it refers to the fact that, no matter how hard agents, producers, studio moguls, and consultants try to predict what viewing audiences will pay money to see, you just never know.

I suspect this saying also applies to science, physics, knowledge, and our understanding of the world around us. Nobody knows anything, at least not 100 percent for sure hands down, guaranteed, without a doubt. We can only theorize about who we are, why we are here, and what will happen when we die. We take the evidence before us, and try to mold it into some semblance of structure, reality, and truth. But we keep forgetting that truth is a matter of perspective . . . and perception. We also forget that everyone has an opinion.

This book is in no way intended to be any kind of last word, rock bottom final truth. Nobody can write a book like that, not even Einstein, so all you skeptics lighten up. Nor am

I trying to compete with people with Ph.D.s. I am an idea person. I leave the math to others far more capable. I can barely balance my checkbook.

This is a book of theories, of possibilities, concepts, dreams, and speculations, based upon a lot of hard (and soft) evidence and a lot of reaching beyond the boundaries of the known into that gray zone, where, let's face it, anything is possible. It is exciting and fun and intriguing and just might resonate with others out there who have had unusual experiences that couldn't be duplicated in a lab or reduced to a math equation. It is one truth in a multiverse of truths waiting to be discovered, if that is at all possible. There isn't a scientist, skeptic, student, or sage alive who knows the ultimate reality of reality. All we can do is guess, and hope that maybe, when we leave this earth, we might get a glimpse of truth. But that's what makes it so exciting . . . the guessing. The hope.

Nobody knows anything; well . . . nobody except whatever created everything. If, that is, there is such a thing as "whatever."

Just buckle up and enjoy the ride.

—Marie D. Jones

One Step Beyond
the Outer Limits
of the Twilight Zone

When science begins the study of non-physical phenomena,
it will make more progress in one decade than in all the
previous centuries of its existence.

—Nikola Tesla

When I first got my hands on Lynne McTaggart's astounding
book, *The Field: The Quest for the Secret Force of the Universe*,
I felt my toes curl. To be honest, I felt everything curl as I read
this stunning book about recent scientific discoveries that
point to the existence of a unifying Force that binds, creates,
and unifies all things. As a student of science, I was thrilled.
But what really got my goose cooking was the author's mention
of how this new discovery of the "Zero Point Field" might be
able to explain something that fascinated me far more than
quantum physics . . . the paranormal.

McTaggart's book about subatomic matter, wave interference patterns, collapsing wave functions, parallel worlds sitting
a hair's breath away from our own, and a field of nothing and
everything that links it all together was not my first exposure
to the idea that there was a purely scientific basis for spooky

and mysterious events that have been reported from time immemorial. I first read about this link in books by "old school" paranormal researchers (and sometimes scientists) such as Jacques Vallee, Jerome Clark, and others who made their mark studying UFOs, alien contact, strange creatures, psychic abilities, poltergeists, and other things that go bump in the night.

Imagine my surprise when books began mysteriously falling into my lap (synchronicity, or some deeper force at play?) solidifying the theory that the world of physics and the world of metaphysics were more closely entwined than ever imagined. Books, such as Dr. Wayne Dyer's *The Power of Intention*, use the concept of the Zero Point Field as the place of pure potentiality, where anything can be manifested into reality. This concept is echoed in books by Deepak Chopra. *Power vs. Force*, the jaw-dropping book by David R. Hawkins, examines the hidden determinants of human behavior and social trends. There was also Malcolm Gladwell's *The Tipping Point*, and Phillip Ball's *Critical Mass*, which dissect common experience and collective consciousness that lead to trends and cycles, and dozens of books I devoured about ghosts, UFOs, strange cryptozoological creatures, poltergeists, and haunting visions . . . all of which carried a strikingly similar theme of *a basis for their existence grounded in real science.*

As an avid reader, student of many subjects, and book reviewer for several popular Web sites, books come my way by the boxful. So I found it a bit unsettling that books with such similar themes kept showing up, begging to be read. Even books I figured couldn't possibly have *anything* to do with this new link between science and the paranormal ended up having *everything* to do with science and the paranormal. These books turned out to be just more arrows pointing me directly toward the most cutting-edge theories of physics, including the multiverse theory, the Zero Point Field, and the quantum proof for inter-dimensional travel and contact between various levels of reality.

I wanted to know what the scientists thought. You know, the skeptical hard-cases who think in terms of mathematical equations and refuse to believe in anything they can't recreate in a lab. The men and women who look at the world through the lens of a device meant to pierce the heart of reality and read the mind of God, the Master Architect. I loved and respected and admired these people, and thought for sure that if anyone could find out what was behind everything else, they could.

And they didn't disappoint me, as I began to read books such as *Parallel Worlds: A Journey Through Creation, Higher Dimensions, and the Future of the Cosmos* by the brilliant physicist Michio Kaku. Or Michael Talbot's *Holographic Universe*, which presents the theory of a universe that is beyond three dimensions advanced by noted physicist David Bohm and neurophysiologist Karl Pribram. Okay, these were the big dogs, the most brilliant minds on the planet, and they believed in the same possibility—that the things we describe as paranormal or supernatural may indeed have a perfectly scientific explanation. Could these events be coming from other worlds, crossing breaches between dimensions, traversing realities, passing through alternate universes, and finally coming to rest at a truck stop near you in the form of a UFO sighting, a ghostly encounter, or a spectral visit from a wolf-like creature standing in your hallway?

This is coming from scientists. MENSA material. Brainiacs. Intellectuals. *People with Ph.D.s!* What is going on here?

I have, since childhood, been fascinated with all things paranormal. I have, also since childhood, been a science fanatic. But never have I been more excited to be alive than I am now, after reading these books, after making this connection. Could modern, cutting-edge quantum physics be on the verge of finally explaining the unexplainable? Had it already done so, and we were all just waiting to catch up? Was the psychic about to merge with the physic?

Was science about to spawn a whole new category known as "*psi*ence?" Were crazy, kooky UFO buffs and far-out psychics about to find themselves in agreement with Harvard physicists with Nobel prizes and fancy titles? Well, that was maybe a bit extreme.

And then it happened. My gardener was edging the red apple on the side of my home when a rock flew up and shattered the entire panel of an upper hall window into a thousand bits of crackling glass. As I stood there staring at the broken window, my first reaction was one of fear. No, make that dread. *Now the bees will come in.* I hate bees with a passion.

Then it hit me. We live our entire lives behind walls and windows that allow us only a passing glimpse of what lies outside our usual "reality." One day, we get a broken window, and suddenly we become exposed to everything that lies beyond, everything that might want to pass through and invade our narrow, safe perception of life, such as bees.

The paranormal was like a swarm of bees. And the world of quantum physics provided those bees with the very thing they needed to enter our reality anytime they chose to. A broken window. A door ajar. An open portal through which anything and everything could travel from one part of the space/time continuum to another. I had found a missing piece to a mysterious jigsaw puzzle that could, when completed, change everything about how we view our world and our place in it. But even more fascinating and exciting, I had discovered *that human consciousness itself* is the key to our reality, and all that it encompasses. That includes the known, and the unknown.

As history shows us, the window has always been broken, for paranormal events have been around since recorded time began. And like Humpty Dumpty, the window cannot be put back together again.

By the way, the following week, my son discovered another broken window on our lower level, only this time there was no visible explanation for how it cracked. After living in my home for almost five years with not so much as a crack to any of my windows, I get two broken windows in one week? Coincidence? I don't think so.

As that cute little girl once said in the movie *Poltergeist*— "THEY'RE HERE!"

The Paranormal

*The object just appeared . . . then it blinked out. I don't
mean it moved away. I mean it just blinked out, like it was
never there at all . . .*

*I looked up and saw this thing in my hallway, coming
towards me. Its eyes glowed. I couldn't move . . . then it
seemed to fizzle out and I heard static and it was gone.*

*I saw a woman move across the room and disappear into
the wall . . .*

*The craft was hovering and rocking back and forth. It
glowed and dimmed, and it stunk of sulfur. Then it got
really bright and vanished into thin air.*

These are typical reports from people who claim to have
experienced a paranormal event. They are gleaned from actual
cases, numbering in the thousands, of material objects and

spectral visions that seem to manifest out of thin air, then vanish back into nothingness. Often they are accompanied by static, popping electrical charges, the smell of burning metal, sulfur, or even sweet perfume. In many events, electromagnetic energy in the immediate vicinity is directly affected, causing problems for nearby power stations, computers, radio towers, aircraft, and even automobiles.

We, as rational human beings, find it hard enough to believe that paranormal beings, UFOs, ghosts, and strange, mythological creatures exist right here in our own universe. Many scientists insist that it is impossible for these events to occur, based upon the known laws of our universe.

But what about the unknown laws of other universes? Imagine the possibility that paranormal events and beings come from somewhere else completely— another dimension, an alternate universe, or a field of pure potentiality existing of nothing but unmanifested energy, entering through the open door of our conscious reality . . . a place where the rules of our world simply do not apply.

In a stunning 2005 poll taken by the Scottish Paranormal Organisation, more people are likely to believe in ghosts and the paranormal than have faith in any organized religion. And another Gallup survey, taken in June of 2005, showed that about three in four Americans profess at least one paranormal belief. That's an awful lot of people who believe in ghosts, UFOs, aliens, ESP, and other phenomena that defy the known laws of science.

The paranormal has existed alongside the normal since the dawn of time, when cavemen and cavewomen saw magic everywhere, infused in the mysterious and often terrifying workings of nature. To ancient peoples, life itself was filled with dynamic and strange wonders that had no explanation other than what the creative mind could conceive. Science didn't exist, physics had yet to be invented, and the laws of nature had no method to their madness. Chaos was the order of the day.

Today, as we cross into the new millennium, certain we are evolving and progressing toward total control of our environment, there are still so many things we have yet to learn, so much that eludes us.

The word *paranormal* means, simply, beyond normal, and includes any phenomenon that can't be easily explained with a known law of science, or easily duplicated in a university research lab. These include angels and aliens, spaceships from other worlds, ghosts and mysterious creatures, ESP and psychokinesis, and remote viewing—places where the laws of physics are nullified and the center no longer holds. The range of phenomena includes everything from the mind-bending mental abilities studied by parapsychologists, to the awe-inspiring physical appearances of alien craft and crop circles, to the shadowy gray zone in between, where objects materialize out of thin air, and mischievous entities wreak havoc on human sensibilities.

But just because these phenomena occur beyond the range of normal experience and often defy scientific explanation doesn't mean they don't exist. And just because the basic law of cause and effect is tossed on its head by these events doesn't mean they can't possibly have their origin in the real world, the world of the normal. The world of science.

In fact, as the following chapters will explain, the paranormal and science may just be two ends of the same yardstick, connected in ways that are only beginning to be explored and understood.

But before we go down that rabbit hole, we first need a basic understanding of the phenomena described as "paranormal."

Fermi Asked First:
UFOs and the Search for Intelligent Life

Perhaps they have always been here. On earth. With us.
— Jacques Vallee, *Revelations*

We are not alone.
— *CE3K* movie poster

It's Venus.
— My dad

Unidentified flying objects and aliens from outer space make up the mythology of our modern era with movies, television shows, and books, documenting both real and fictional accounts of human interaction with entities from other worlds. Google "UFOs" on the Internet and the search results turn up millions of Web pages devoted to the subject.

We are obsessed with little green, excuse me, gray, men from planets far beyond our own simple solar system. ET has entered our collective consciousness, and refused to let go.

But our obsession is nothing new. It began, in fact, when we as a species began. From ancient cave paintings depicting "astronauts" wearing helmet-like objects, to the more modern waves of sightings all across the globe, the mantra of "we are not alone" permeates our existence similar to some mythical archetype, calling us to expand our vision of the universe . . . and beyond.

This book in no way intends to cover the entire UFO phenomenon. No author could possibly cram the amazing history of ufology into one book, let alone one *chapter*.

Still, a little background won't hurt.

The Early Years of UFOs

The modern era of the UFO enigma began in the 1940s, when several key events took place. UFO sightings had peppered the news even before then, including the notorious "foo fighters" during World War II, but what happened in the latter half of this decade set the stage for the study of all things extraterrestrial for years to come.

In June of 1947, a pilot named Kenneth Arnold unwittingly set off the modern era of UFO sightings when he spotted an unusual craft over the Cascade Mountains in Washington. Arnold, working for the Central Air Service at Chehalis, took off for Yakima, but was delayed by a search for a large marine transport that allegedly went down on or near the southwest side of Mt. Rainier. Arnold was searching the ridgelines when he decided to turn around and make another attempt.

At an altitude of 9,200 feet, Arnold spotted a DC-4 about 15 miles away. The skies were crystal clear and a pilot's dream. A few minutes passed and Arnold saw a bright flash reflected off his craft. He looked around and saw a chain of nine aircraft moving north to south at about 9,500 feet. The aircraft were moving rapidly toward the mountain, and Arnold thought it odd that they had no tails like the usual jet planes.

After observing the objects for some time, Arnold became more and more anxious. As a pilot, he was familiar with most aircraft, but these seemed to move in a way he couldn't recognize.

Arnold searched for the downed marine plane for another 15 minutes or so, all the while growing more distressed over the nine mysterious craft.

When Arnold landed at Yakima, he told a good friend about the incident, and eventually shared his story with pilots in Pendleton, Oregon, including former Army pilots who assured him that he was not crazy. Eventually the news spread and Arnold began getting phone calls from all over the world supporting his claims. He even invited the FBI and the Army to investigate, but got no response.

But Arnold made history when he told reporters that the objects he observed acted like "pie plates skipping over the water." Once the media got hold of the story, the phrase "flying saucers" made its debut, possibly first in the *East Oregonian*, the local Pendleton paper. No one can pinpoint for sure who actually coined the phrase, but it was a label that would stick for decades more, until the term "UFO" became the common moniker.

In July of that same year, residents of Roswell, New Mexico, reported a "big glowing object" that raced out of the Southeast skies toward the Northwest. Witnesses described it as oval-shaped, like two inverted saucers placed together. A few days later, on July 8, a man named Walter Haut, who served as the PR officer for the Army Air Base outside of the city, issued a stunning press release that made the newspaper headlines, stating that indeed a "flying disc" had been recovered from a nearby ranch and taken to Roswell Army Air Field.

Once the wire services picked up the story and ran with it in newspapers across the globe, General Roger M. Ramey, commander of the Eighth Air Force District, immediately announced that the earlier report had been in error, and that the object had been only a weather balloon.

Still, rumors persisted of a downed UFO and debris composed of matter not of this earth. The legend grew to include dead alien occupants, hidden away at the mysterious locale, Hangar 18, at Wright-Patterson Air Force Base. So pervasive has the Roswell mystery become in our modern culture, UFO researchers are still convinced the government has been hiding the truth for more than five decades.

Project Blue Book

Throughout the 1950s and into the 1960s, television and motion picture producers milked the public's fascination with alien spaceships. But the government wasn't taking things so lightly, despite constant denial of their interest in the subject; in 1947, the United States Air Force launched Project Blue Book, an official and active investigation into the UFO phenomenon. Project Blue Book evolved from two prior studies, Project Sign and Project Grudge, both initiated by the Air Force.

Project Blue Book, on which noted astronomer Dr. J. Allen Hynek served as consultant, lasted 22 years before it was shut down by the Air Force. But in those two decades, Hynek and other Blue Book investigators examined over 12,600 UFO sightings, 701 of which remained unidentified by other means. The final results of the project stated that no UFO was ever perceived as a threat to national security, that no evidence was discovered to indicate UFOs possessed technological capability beyond our present scientific knowledge, and that there was no evidence that these objects were of extraterrestrial origin.

But many modern UFO researchers feel Blue Book was more of a PR snow job than a real scientific and professional investigation into an enigma that, despite the project's rejection of, was still continuing, and even increasing. Many researchers took offense to Blue Book's conclusions that UFO sightings were the result of mass hysteria, fabrications for purposes of publicity, or simple misidentification of conventional craft (how they passed off the hundreds of sightings by military and professional pilots is beyond me). Decades later, in an interview with ABC News, former Project Blue Book project director Col. Robert Friend would confirm those suspicions saying, "What they wanted to try to do was, I think, to reeducate the public regarding UFOs, to take away the aura of mystery." They did

Imagine waking up to this headline, as many Suffolk, England, residents did one morning.

this by keeping UFOs out of the newspapers, and out of the public eye, by repeating over and over that they were nothing but weather balloons or stars on the horizon. (Although it didn't always work with the media!)

Repetition of disinformation has, and still is, the number one tool of psy-ops, or the psychological operations the government uses to keep the public accepting a certain paradigm or to keep them from questioning a position. Psy-ops is a reality, and with the UFO enigma, one that worked well at keeping the subject matter well below the radar of "serious study."

Ironically, J. Allen Hynek would begin his Project Blue Book involvement as a die-hard skeptic, but end it as a believer, devoting the rest of his life to investigating sightings and calling for serious inquiry into the phenomenon. He claimed that what finally convinced him were the many sightings by military pilots that simply could not be written off as hysteria or human error. Pilots know about stuff in the sky.

Hynek would also invite the ridicule and rejection of many of his fellow scientists. Eventually, in 1973, he founded the Center for UFO Studies in Chicago, and continued his quest for truth until his death in 1986.

Sightings and Speculation Across the World

UFO sightings would sweep every part of the world, including Mexico City, Belgium, and the Hudson Valley UFO wave of 1983, witnessed by over 5,000 residents of the Tri-State (New York, New Jersey, Connecticut) area, including my own grandparents, who had a sighting from the porch of their Bridgeport, Connecticut, home.

UFOs gained even more exposure and credibility when U.S. president Jimmy Carter confessed to his own sighting in Leary, Georgia in 1969 (although President Gerald Ford beat him to the punch, calling for a congressional investigation after a wave of sightings in 1966). Ronald Reagan reported seeing a UFO in 1974 while he was governor of California. Former president Bill Clinton had a strong interest in UFOs, and according to then Associate Attorney General Webster Hubbell, Clinton instructed him to find out if UFOs existed. Clinton also spoke out against the Air Force for its failure to be forthcoming with information about the possible recovery of alien bodies at the Roswell crash site in 1947.

Carl Jung, the Swiss analyst and father of modern psychology, believed UFOs were not of earthly origin and even wrote his own book, *Flying Saucers* in 1959 to toss into the fray. United States senator Barry Goldwater, while in office no less, repeatedly called for serious inquiry into UFOs, labeling the subject "above top-secret." Even the father of modern rocketry, Dr. Hermann Oberth, would express his conviction that UFOs were extraterrestrial vehicles of high technical design.

More than a dozen astronauts, including Gordon Cooper, Buzz Aldrin, Neil Armstrong, Donald Slayton, and Wally Schirra stated they, too, had seen objects while out in space they could not identify. In 2004, Soviet cosmonaut Vladimir Kovalenok told the Moscow press he, too, saw an unidentified craft, but was met with widespread ridicule. Kovalenok's fellow cosmonaut, Viktor Savinykh, also witnessed the event, but the two were unable to photograph the object before it exploded. Kovalenok later learned that specialists had registered considerable radiation emission the day the astronaut saw the object. "I do not believe it when astronauts say they have never seen anything extra-ordinary in space," concluded Kovalenok in the August 16, 2004, issue of *Pravda*.

Despite ongoing media ridicule and government denial, even something as simple as the categorization of UFOs by the Center for UFO Studies would become a part of our everyday culture.

> Close Encounters of the First Kind: Visual sighting
> Close Encounters of the Second Kind: Physical effects
> Close Encounters of the Third Kind: Contact and a major
> motion picture contract with Steven Spielberg

Researchers would eventually add a fourth kind: abduction. This was set in motion by the case of Betty and Barney Hill, a couple heading home from southern Canada one night in September of 1961, when they were waylaid by an alien craft while driving through the White Mountain area of New Hampshire. The Hills would eventually report their mysterious otherworldly encounter to the Pease Air Force Base, and were told nothing of note happened at the time, although reports released 20 years later show a UFO observed on the base at the same time.

The Hills' story would become one of the most amazing in UFO history, when ufologist and astronomer Walter Webb would visit the Hills and begin an investigation that would uncover a host of oddities, including "missing time" experienced by the couple. Eventually, Betty and Barney were sent to see Dr. Benjamin Simon, a psychiatrist in the Boston area. Simon regressed the couple by hypnosis, allowing their subconscious minds to then release the events of the night that their conscious minds were blocking. The Hills reported, under great duress, that they had experienced a classic UFO abduction by small beings with large cat-like eyes and pale white skin. Betty underwent awful tests by the aliens, while Barney had semen extracted. Betty then reported being taken on a tour of the spacecraft. She was given a star map of the Zeta Reticuli system, from whence the little critters had come.

The Hill case now has its place in the UFO Hall of Fame, spawning movies, books, and the continuing debate over alien abductions and the use of hypnosis to extricate memories so traumatic, the conscious mind simply can not handle them. Throughout the next three decades, UFO abduction scenarios would become almost commonplace, changing the face of the phenomenon forever. Some of the more famous cases include the case of Betty Andreasson, who claimed abduction in January of 1967; the Hickson/Parker abduction case in Pascagoula, Mississippi; the Snowflake, Arizona, Travis Walton abduction that spawned a movie called *Fire in the Sky*; the chilling experience of one Linda Cortile on the Brooklyn Bridge in November of 1989, which spawned the book *Witnessed* by noted UFO researcher Budd Hopkins; and the incredible visitations that occurred over a period of time to a woman named Kathie Davis in Copley Woods, Indianapolis, covered in detail in Hopkins' best-seller, *Intruders*, which was also made into a TV mini-series. And these are just the tip of the iceberg.

Meanwhile, the scientific world would launch its own attempted investigation into the possibility of extraterrestrial life, with SETI, the Search for ExtraTerrestrial Intelligence. This major project was inspired in 1959 when two Cornell University physicists, Giuseppe Cocconi and Philip Morrison, published an article in *Nature* examining the potential of microwave radio to detect signals from the stars.

Based at NASA's Ames Research Center and the Jet Propulsion Lab in Pasadena, California, SETI would sweep the stars, examining 1,000 sun-like stars in a Targeted Search, capable of detecting weak and sporadic signals in a Sky Survey that covered every direction.

In the early 1990s, the U.S. Congress would terminate federal funding of SETI, but the search continues with public and private donations as the mainstay for keeping the program going. Project Phoenix, a more comprehensive Targeted Search, was also launched in 1995 with the hopes of taking advantage of a historical opportunity to search the skies before radio interference from terrestrial sources threatens to drown out any hopes of picking up a weak signal from somewhere beyond.

But the questions remains . . . beyond *where?*

While doing research at the National Radio Astronomy Observatory in Green Bank, West Virginia, astronomer Frank Drake discovered a way to mathematically estimate the number of worlds that might harbor beings in the Milky Way galaxy with technology sufficient to communicate across interstellar space. The Drake Equation was formulated in 1961 and is as follows:

$$N = R^* \; fp \; ne \; fl \; fi \; fc \; fL$$

N is the number of civilizations in our galaxy with which we might expect to be able to communicate at any given time.
R* is the rate of star formation in our galaxy.
fp is the fraction of those stars that have planets.
ne is the average number of planets that can potentially support life per star that has planets.
fl is the fraction of the above that actually go on to develop life.
fi is the fraction of the above that actually go on to develop intelligent life.
fc is the fraction of the above that are willing and able to communicate.
L is the expected lifetime of such a civilization.

Frank Drake's own original solution to the Drake Equation estimated 10,000 communicative civilizations in the Milky Way. However, these figures were destined to change as we learned more and more about the sheer massive size of our universe, which may contain an infinite number of galaxies . . . and that's just OUR universe! Only time, and further exploration, will tell. To see a photo of Drake, go to *www.setileague.org/photos/aaas/drake.gif*.

The Search for Proof of Life

My own connection to UFOs began as a child. Obsessed with the stars and planets, I wrote a book at the age of five called *Life on Mars*. No agent would take it on, perhaps because it was written in crayon and the cardboard cover was tied together with shoestrings, but I held out hope. My father, a geophysicist, enjoyed talking about the subject with a neighbor who was a UFO buff. We had UFO books around the house, and, although his background was hard science, my dad was wise enough to consider the existence of extraterrestrial intelligence.

But his science background also meant he was the go-to guy whenever a strange object appeared in the skies over Garnerville, New York. I remember one particular night when a bright object in the night sky had my entire neighborhood in an uproar, and my dad, the spoiler, shot down our dreams of alien visitation and international fame with just two simple words—"It's Venus." We all went back to what we were doing, the kids playing flashlight tag and adults visiting with neighbors, collectively feeling as though someone had just popped our favorite party balloon. Oh well.

After later relocating to California, I began working with local MUFON (Mutual UFO Network) groups and joined the Center for UFO Studies. I passed field investigator training and worked with a team on local sightings both in northern San Diego and Los Angeles. Many of the cases we worked on could be easily passed off as hoaxes, many more as misidentifications of known objects such as small planes, satellites, weather balloons, and aircraft, which were plentiful around the military bases and airports that pepper the northern area of San Diego County. Some were downright weird and smacked of events occurring in the minds of the witnesses, rather than as an outside, physical manifestation. Others sounded like the run-of-the-mill sightings that other groups were reporting all over the world: "Bright light in sky. Traveled at high speed. Creeped me out."

But the cases that always intrigued me involved witnesses who repeatedly referred to objects that "appeared instantaneously, then disappeared as if into thin air," suggesting that some UFOs just might be coming from, and returning to, a place beyond our physical universe. This was especially true when the witnesses were police officers, pilots, and engineers not usually prone to making up stories about flying pie tins. In fact, in my humble opinion, the more credible the witness, the more hesitant he or she would be to even report a sighting in the first place.

In general, ufologists agree that these phenomena can be explained as having one of the following points of origin:

- Our earth, as in hollow earth theory or from deep in the ocean
- Our world, as in top-secret experimental craft made by the military or government
- Our known universe, as in other galaxies or the outermost corners of our own universe
- Other universes
- Other dimensions
- And strangest of all, from our own consciousness

Not surprisingly, it would be far easier for researchers (and the public) to focus on the first three. It is a little easier to suggest we are being visited by creatures from other planets or from within the bowels of our own earth than to even contemplate breaches in the space/time continuum. It's even easier for us to believe our own government is hiding its involvement in the UFO enigma, especially in light of the abundance of evidence that it has been doing so since before World War II.

In their fascinating book, *Strange Secrets*, authors and investigators Nick Redfern and Andy Roberts document the involvement of governments on both sides of the globe in the UFO phenomenon. Using official files, the authors claim that the United States authorities not only knew of the existence of UFOs, but in some cases may have known what they were. In fact, declassified documentation exists proving that our government was deeply involved in the building and testing of prototypes for typical UFOs at top-secret labs across the country.

The Avrocar and Other Prototypes

One such man-made flying saucer was the Avrocar, developed at the Avro-Canada plant in Malton, Ontario, in 1953. This circular craft sucked in air from the top, expelling it at the edges of the disk, and flew at altitudes of 5 to 6 feet. But higher altitudes seemed out of reach due to instability, and the project was abandoned. The prototype now sits in an Air Force Museum in Fort Estis, Virginia. Many ufologists felt that the cute little Avrocar, and its failure to rise to the stars, was just enough to keep public attention away from the truth, that governments and military agencies took UFOs and the potential to master their dynamics of flight very seriously.

The Avrocar project barely got off the ground before being shut down.

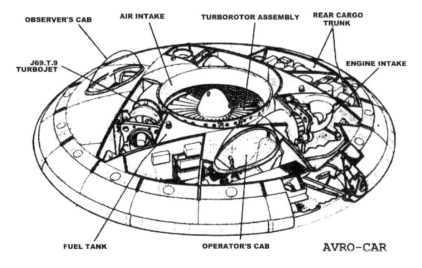

OBSERVER'S CAB AIR INTAKE TURBOROTOR ASSEMBLY REAR CARGO TRUNK

J69.T.9 TURBOJET ENGINE INTAKE

FUEL TANK OPERATOR'S CAB AVRO-CAR

Other prototypes were created, tested, and eventually abandoned, launching a competition between countries such as the United States, the United Kingdom, Canada, and the Soviet Union. German scientist Richard Mehta was considered the "father of saucerology" for his technological expertise. Mehta was hired by the German Air Force to build a saucershaped craft that could ascent vertically and shoot down allied planes, but World War II ended before his ship got off the ground, so to speak. The craft was given the code name "Silverbug," and research continued on saucer-type flying machines even after the war via the notorious Project Paperclip, the ultra-classified recruitment of Nazi scientists onto American soil.

Many ufologists and military aircraft buffs claim that research on homemade UFOs continued at spooky classified facilities all over the world for decades, many in the good ol' U.S. of A, referring to the many reports of UFOs seen near and around such "secret" and not-so-secret bases. Researchers Nick Redfern and Andy Roberts even contend that the high volume of UFO sightings near military installations might in fact be because the military itself was operating the craft, citing a number of top-secret insider reports discussing the building and testing of various homemade flying saucers.

The Government and UFOs

Other declassified documents show a serious interest in the phenomenon on the part of the FBI, CIA, Air Force, NSA, and British Intelligence, including a revelatory document from the NSA titled "UFO Hypothesis and Survival Questions" that examines the various theories of origin and how they might affect humankind (especially in light of national security). Interestingly, this document suggests that not all UFOs are hoaxes, and rules out hallucinations as a main cause for worldwide sightings. The serious concerns include whether or not UFOs are of natural origin, which called into question the capability of air warning systems to correctly diagnose an attack situation from, say, a flock of geese; UFOs as secret earth projects, which was already a serious issue facing competing countries in cases of developing advanced technology; UFOs as intraterrestrial intelligence, which then raised the questions of how advanced alien technologies might perceive us: as a potential threat or as inferior creatures that needed to be conquered!

Though they would continue to officially deny any knowledge of involvement with UFOs, the government and military authorities of the United States, the United Kingdom, and the former Soviet Union all extended their own research, investigation and, often, suppression of evidence to other paranormal events such as crop circles, cattle mutilations, and even the Loch Ness Monster! Tin foil hats aside, it makes sense that if a strange phenomenon is happening to the populace, the government in charge would want to know about it and ultimately get the upper hand.

But even our most elaborate high-tech labs and top-secret bases couldn't possibly construct objects that vanished into nothingness, and appeared and disappeared before numerous witnesses as if they had total disregard for our known laws of physics. We might be building V-wings and Blackbirds and even mushroom-shaped hovering crafts, but they all have to follow the laws of physics that governed matter and energy.

Perhaps objects that came from elsewhere would have their own laws of physics.

The Science of Sightings

Dozens of books and now hundreds of Web sites document UFO sightings, and when one has time to examine many of them (aside from obvious hoaxes and cases of misidentified known craft) two trends seem readily apparent.

UFOs are coming from and returning to some point in known space.

UFOs are coming from and returning to some point beyond space/time as we know it.

Take for example this sighting in Chateauguay, Canada, found on the National UFO Reporting Center's online sighting database. The shape was changing; the duration was between one and two minutes. Big deal. But the object's description begs a bit more attention. *"Pulsating bright greenish white object that hovered and then disappeared into thin air."*

A few more 2005 sightings echo that description:

. . . black object moving slowly in broad daylight suddenly disappears . . .
—Arden, California

The object came out of nowhere, did a dance, and then disappeared.
—Waukesha, Wisconsin

Came from what? Disappeared into what? And these simple sighting reports are rampant throughout UFO literature, many coming from specialists such as pilots, astronauts, military officials, and ordinary people. Clearly these vehicles were originating from places other than distant planets or solar systems.

No, something else was going on that went beyond the concept of interstellar space travel—something that involved dynamics not seen or experienced in our own everyday reality—something that whispered of the ability of those who were perpetrating the events to manipulate space and warp time and, quite simply, "disappear into thin air." Something that either our government knew about and didn't want to share, or couldn't quite understand and didn't want to admit.

There are two reasons why any government would cover up the UFO phenomenon (or cover up anything for that matter):

1. They know and don't want us to know they know
2. They don't know and don't want us to know they don't know

The second option is more frightening, especially in light of today's concerns with terrorism and homeland security. Imagine the homeland *insecurity* that might arise if the public thought we could be dropped in on at any time by funny looking little grays. It's one thing to hide the truth from the public because you don't want the enemy to figure out how it's being done (as in military maneuvers or wartime planning) or because you want to control any potential new energy source that might be back-engineered from a crashed UFO (as in "stop them from cutting into our profit margin"). It's quite another to hide the truth because you don't want anyone, including the public, to know just how vulnerable we all are to becoming Tom Cruise in *War of the Worlds*. (Had I said Gene Barry, half of you would have said, "Who?")

If validation of the UFO phenomenon isn't going to be forthcoming from our authority figures, and because many serious ufologists are unfortunately stereotyped as single-minded believers who know little about science, the public is left with the media, namely the entertainment industry, as the sole source of a possible answer to the UFO question. No wonder the public mindset has become centered on the most sensational aspects of the phenomena, such as the growing number of abduction claims and the contactee movement where ordinary Joes and Janes report ongoing and often prophetic conversations with aliens named Rantu and Gorgon.

As one of the few scientists who does take the UFO phenom seriously, Jacques Vallee stands out. He wasn't the first to do so by any means, but the astrophysicist and computer scientist was the most notable scientist-turned-UFO researcher to talk widely about the possibility that UFOs might indeed be "not of this world." Only Vallee (author of such groundbreaking books as *Dimensions: A Casebook of Alien Contact* and *Revelations* and the model for the French scientists in Steven Spielberg's *Close Encounters of the Third Kind*), didn't mean not of *this* world in the same sense as other scientists who had dared enter the fray were thinking of. Vallee, who spent 30 years looking into every possible claim and theory, came to the startling conclusion that many UFOs might be coming from other dimensions in the space/time continuum.

Vallee's extensive research also delves into a subject that is thought of by many by-the-book researches as taboo—the role of the spiritual quest in understanding the mythology of UFOs and how, quite possibly, human consciousness is involved in the manifestation of the enigma.

Heavy-duty stuff for a scientist to be delving into.

Vallee set forth a series of perspectives that takes ufology to the next level, even claiming that UFOs play a powerful and particular role in the history of human affairs, altering and shaping paradigms, and manifesting in accordance with the collective consciousness of the time. These ideas boggle minds already trying to figure out how UFOs manipulate electromagnetic fields, what types of propulsion systems they might utilize, how they get beyond the limitations of light speed and gravitation, and even why they seem to come in cycles and waves. But even Vallee was not immune to the basics. He believed there was a cover-up, that science was being suppressed, and that some UFOs could indeed be of this world. He also investigated many claims of objects (and occupants) that behaved more like holographic projections, such as these two from *Revelations*:

> June 1962, Verona, Italy—*Following a UFO observation, a woman was awakened by a feeling of intense cold and saw a being with a bald head near the house. She called other witnesses, and all saw the apparition shrink and vanish on the spot "like a TV image when the set is turned off."*
> November 1968, France—*A doctor saw two large disk-shaped objects merge into one, and the single object sent a beam of light in his direction. It vanished with a sort of explosion, leaving a cloud that dissipated slowly.*

These cases intrigued Vallee, for they hinted at origins far more complex than the usual extraterrestrial, yet interstellar, playing field. He examined dozens of cases that led him to believe most witnesses were not seeing a material object at all, but rather a light of impressive, and often hypnotic proportions. After the examinations, Vallee questioned modern science's ability to deal with a phenomenon that included effects of electromagnetic radiations, not just on things such as cars, power plants, and radar screens, but *on the human brain and nervous system*. At the time (in 1988) there was virtually no scientific research available.

Other scientists would be drawn to the more scientific and lab-worthy aspects of UFO studies, namely the physical effects and material traces, often involving electromagnetic fields. By focusing on the physical effects, which often provide researchers with important clues as to possible propulsion methods and added credibility to the phenomena as being something other than a hallucination or mental aberration (it's hard to hallucinate traces of radiation or burned treetops as a result of a low-flying craft), serious investigators could get one step closer to finding tangible proof of alien craft. This was especially important at alleged UFO landing sites and special cases involving animal effects, damage to surrounding areas due to high heat (charred brush or downed trees), trace remains of magnetized metal, lightning-like effects in soil, and those involving power outages, disruption of electronics, and automobile and other vehicular malfunctions.

Fermi's Paradox and Other Theories

Perhaps the first serious scientific study of UFOs was Project Identification, which began in 1973 under the guidance of Harley Rutledge, Ph.D., whose specialty was solid-state physics. Rutledge would log several hundred hours of observation time in this field study, monitoring phenomena in real-time. Calculations were made for object velocity, course, position, distance, and size, and the final results were documented in the 1981 book, *Project Identification: The First Scientific Study of UFO Phenomena*. Rutledge's observations would conclude that some UFOs are "pseudo stars"—stationary lights camouflaged by constellations; some were clearly mimicking known aircraft; some were violating all known laws of physics; and, stranger still, some objects responded to lights being switched on and off, and to verbal and radio messages. But the strangest find of all was that at least 32 recorded objects were lights that moved in synchronized motion with the actions of the observers.

Several physicists, such as Stanton Friedman, who was employed for 14 years as a nuclear physicist for major companies working on advanced and often classified projects such as nuclear aircraft and fission and fusion rockets, would become well-known proponents of the "flying saucers are real" stand, using their credentials and technical expertise to propel the study into more serious territory.

Other respected mainstream scientists such as James E. McDonald, Peter Sturrock, and Auguste Meessen would argue that UFOs deserved further study, including case-by-case analyses using the scientific method. Dozens of men and women of science, experts in various fields of technology, and even top psychologists would enter the UFO debate in the years to come, creating a much more credible place from which to perform and present research. This included Bruce S. Maccabee, Ph.D., a physicist and active UFO researcher since the 1960s, who would bring years of experience to the Naval Surface Warfare Center, working on optical data processing, generation of underwater sound with lasers, and even strategic missile defense using high-powered lasers.

Maccabee was joined by physicists James Deardorff, Harold Puthoff (who we shall hear more about later in the chapter on the Zero Point Field), and astrophysicist Bernard Haisch in a joint proposal in the January/February 2005 issue of the *Journal of the British Interplanetary Society*. This proposal suggests that, according to a January 14, 2005, story on Space.com, it is a mistake to reject all UFO reports "since some evidence for the theoretically-predicted extra-terrestrial visitors might just be found."

The men were referring to Fermi's Paradox, originated by physicist Enrico Fermi who once, over lunch with fellow physicists discussing UFOs, asked the question, "Where are they?" That simple question led to debates about the probability of planets that could, and would, give rise to advanced civilizations capable of interstellar travel.

Here were a bunch of brilliant scientists actually talking, without fear of ridicule, about the possibility, no, the *likelihood*, of ET visitors being out there, and maybe even coming here.

Deardorff, Puthoff, and Maccabee's proposal would involve the latest discoveries in cutting-edge physics, theories such as wormholes, superstrings, inflation, and other dimensions in the space/time continuum, all of which we will delve into in Parts II and III of this book.

A comment by Deardorff to the press would set the stage for things, exciting things, to come: "It would take some humility for the scientific community to suspend its judgment and take at least some of the high quality (UFO) reports seriously enough to investigate . . . but I hope we can bring ourselves to do that."

And if that wasn't enough to get the UFO crowd in a joyful frenzy, a man by the name of Michio Kaku, who Vallee mentions as a maverick physicist to watch out for in his 1991 book *Revelations*, would write several stunning books and come up with a few nifty theories that would create incredible new possibilities for the potential existence, and possible travel methods, of UFOs.

But before we go off in search of strings and things, a spirited discussion beckons . . .

Spectral Spectacles: Ghosts, Poltergeists, and Things That Go Bump in the Night . . . and Day

All argument is against it, but all belief is for it.
> —Samuel Johnson, on the existence of ghosts

Something you cannot explain to another person is called nistar—"hidden," like the taste of food, which is impossible to describe to one who has never tasted it.
> —The Essential Kabbalah: The Heart of Jewish Mysticism

It all began with a ghost.

My entry into the world of ghosts began as a child with an old, tattered copy of a Nancy Drew novel (an original printing mind you, now worth a lot of bucks on eBay had my parents not sold them for pittance in a garage sale), in which our plucky heroine found herself once again knee deep in some spooky mystery complete with ghosts, moving objects, and things that were not what they seemed to be.

It wouldn't be until my teenage years when I would have my first real experience with a ghost while staying at a friend's house. Her family had been plagued with poltergeist-like activity for about a year, and I reluctantly agreed to stay overnight, despite stories of flying objects, red flashing eyes in the hallway, sensations of heat or cold, and other creepy occurrences that convinced me, no matter how badly I had to go, to not leave my friend's room at all that night for the bathroom, which was down the hall.

The next morning, I was relieved (in more ways than one!) that I had made it through alive, dry, and unscathed, until my friend and I were sitting in the living room watching the old black-and-white version of *Auntie Mame*, and I distinctly felt a cold wave of energy move through me. Just as my body tightened in unexpected terror, my friend, who was sitting on the couch behind me (I was on the floor eating potato chips), said, "There. Did you feel it?"

Up until that point, I had been convinced that my friend, and her family, all of whom were very strict and devout Lutherans, were imagining or hallucinating these events. Each family member seemed so sincere, and they struggled to find the cause of the mischief, even getting rid of antiques one by one, sure they were vessels for the captured energy of spirits. We all wondered about possession, but thought that more difficult to believe in than ghosts. There was a teenage sister in the house, and we did read about cases of poltergeists that preyed upon young girls. That idea was shot down, though, when the sister moved out of the house and the chaos continued.

But from that morning on, after truly feeling something move across my body that could not otherwise be explained (all doors and windows were shut, heater not on, AC not on, and so on), I believed. Shortly afterward, I began to study how to do an exorcism, especially once I was told that the family's church would not do anything about it. I went to the library and got every book I could find (this was before the Internet) on exorcisms and demons. After about six months of reading and readying myself to single-handedly try and purge their house of demonic forces, I chickened out after a very creepy experience in my bedroom while attempting to invoke demons while in a circle I had charged with positive energy. I was clearly playing with fire and abandoned my efforts to help my friend exorcise the demons.

Instead, I encouraged them to ditch every antique they had.

Ultimately, my friend and her family moved away. We lost touch over the years, but the few times we talked, I never heard another word about moving objects and mischievous ghosts. Maybe it all ended when they left that house (it was a really dark and spooky house). But maybe not. The new owners of their home had it razed and rebuilt from the ground up. I always wondered why.

Everybody knows someone who knows someone who has seen a ghost. I did a few experiments recently to prove how widespread this phenomenon truly is. Each time I sat at a table with a group of people, I found a way to bring up the subject of ghosts. I am always amazed at how many stories come forth, often from people who had, up until then, "told no-one outside my family." My own grandmother, a devout Catholic, had three stories of her own involving distant aunts and cousins, many in her native Italy. One was especially chilling, involving a toddler who refused to go upstairs in a particular two-family house. The baby would cry and wail and thrash at the bottom of the steps. It was later learned that someone had been killed in the upstairs apartment.

Others I talked to reported encounters at famous places known for ghostly activity, such as my hometown of San Diego's Whaley House and Hotel Del Coronado, both of which harbor a history of hauntings that continue to draw tourists from all over the world. I have never been to the Whaley House, but I can tell you from experience there are parts of the Hotel Del where you can actually feel the energy level shift as the air literally becomes oppressive.

Real Life Ghost Stories

Across the globe, legends abound about spirits who hang out and haunt tourists and locales ripe with paranormal activity. Many of these places are documented in books such as Jeff Belanger's *Encyclopedia of Haunted Places: Ghostly Locales From Around the World*. Belanger has investigated each locale he writes about, and approaches the phenomenon with the objectiveness of a reporter. He also runs a Web site called Ghostvillage.com, which serves as a forum for ghost sightings all over the world, sent in by people who want their story documented or want to share and exchange ideas and information on what ghosts might be, and where they come from.

Ghostvillage.com is one of hundreds of Web sites devoted to all things ghostly. About.com has the most comprehensive paranormal Web site devoted to all things spooky (*http://paranormal.about.com*), thanks to the amazing work of About Guide Stephen Wagner. From the dawn of recorded history, ghost stories have appeared in religious texts, fiction, and literature, and even manuals on science and philosophy. And they have often made the headlines of newspapers, such as the January 2005 issue of the *Heritage Scotsman*, which reported the "Tourist-Terrorising Mackenzie Poltergeist."

Reporter Diane Maclean documented the creepy cases of witnesses in the Greyfriars Kirkyard area who lived near a graveyard with a history of executions and murder at the hands of a 17th century judge and lord advocate, Sir George Mackenzie. It seems this was one nasty guy, imprisoning thousands of Covenanters, a powerful Presbyterian political force in 17th century Scotland, and even displaying some of their heads around prison walls (once removed from their bodies, that is). Others were starved and tortured at the whims of the mocking and hateful Mackenzie.

Then, in 1999, a homeless man was caught sleeping in the mausoleum of the mean Mackenzie, and the locals claim this disturbance of the coffin was the trigger that awoke the ghost and set off an ongoing series of poltergeist-like disturbances. More than 350 documented attacks have occurred, and of those, 170 people were said to have collapsed at the site. Tourists report hot spots, cold spots, and every kind of spot in between. Worse still, many are bloodied, bruised, pushed, and poked by this unseen visitor of the Black Mausoleum. One local, who was documenting the cases over time, even claims the ghost burnt down his house, with all of his documents inside. Consider this a warning to anyone contemplating a snooze at the local cemetery!

There is no dearth of ghost stories in our modern media. During the writing of this book, there were no less than four network series on television involving ghosts and communication with the dead, and that doesn't include cable. You cannot get away from ghosts, as I discovered when I opened the December 12, 2005, issue of the *Hollywood Reporter*, to find a story about the creator/producer of the popular series *Ghost Whisperer*, John Gray. He told the *Reporter's* Kimberly Speight about his own ghostly encounter at his New York home. Gray brought in real-life medium Mary Ann Winkowski, the woman the series is based upon, to help the ghost "cross over." Since then, Gray stated, everything went back to normal in the house.

Ghosts, and ghost stories, are everywhere.

Sitting down at a table and asking a question, or visiting a Web site and posting a ghost story are not scientific methods by any means, but there are people out there actively investigating ghosts and other spectral spectacles, and they are doing it with a lot of help from science.

The Ghost Hunters' Theories

"Ghost hunters," as they are often called, are at a distinct disadvantage when it comes to collecting and documenting evidence. Unlike UFOs, which sometimes leave physical traces or effects on their surroundings, ghosts come and go and leave little or no trace, except for the fear and shock (or in the case of the Mackenzie cases, bruises) of the witnesses. Ghosts are difficult to photograph, and often appear as nothing more than a plasma-like orb of light or a moving fog with no distinct shape. Those sightings that do take on a clearer appearance can be just as elusive to capture on film, suggesting that ghosts are some kind of trapped energy we have yet to understand.

Paranormal investigators are convinced that these apparitions, also known as spirits, specters, and phantoms, are real, but opinion varies on what they might actually be. Some think of ghosts as manifestations of energy trapped in our world, but perhaps originating somewhere else. Others see ghosts as the disembodied spirits of the dead, the actual life force of a person aching for rest and closure of a life that ended badly. Others view them as both objective and subjective in nature. In other words, a ghost may take on an objective physical appearance, say as a floating orb of light or as a figure of an old woman striding across the hallway, but also may be something emanating from the mind of the witness, as in poltergeist activity.

The Bell Witch hauntings in Adams Station, Tennessee, may be a little bit of everything. For 200 years, residents of the small farming community have witnessed strange and frightening poltergeist activity, associated with the spirit of a woman who claimed to be a witch. The woman, whom many believe was named Kate Batts, was a neighbor of a man named John Bell. The Bell Witch (who may have been the inspiration for the film *The Blair Witch Project*), lived in 1817. The story has Batts accusing Bell of cheating her out of land, and swearing on her deathbed to get revenge.

The Bell family began experiencing disembodied voices, flying furniture, hair yanking—they even had needles poked into them. Word of the hauntings spread, and people came from hundreds of miles away to see if they could spot the Bell Witch at work. Even General Andrew Jackson gathered a group together to investigate! According to the Martin Van Buren book from 1894, *An Authenticated History of the Famous Bell Witch*, Jackson got his proof and then some when his horse-pulled wagon got stuck on a smooth stretch of road, at which time he heard a distinct "metallic" voice coming from the bushes, warning that he and the ghost would meet again that night. The horses and wagon were thus freed, and Jackson reportedly experienced some physical manhandling, or should I say "witch-handling" that night in the form of slaps, pinches, and hair pulling!

Some locals claim the mischief continues to this day.

Still, others believe ghosts are the emotional memories of a person who died a traumatic death, rather than their actual spirit body. These memories keep the ghost trapped in the same physical location where he or she died, destined to play out the same familiar acts again and again until some form of release dissipates the memory. A more scientific theory suggests ghosts are shadowy images of a real image projected from somewhere else, such as a hologram, which would fit in nicely with the holographic universe theory we will delve into later.

In any event, most parapsychologists and ghost hunters admit that these apparitions and entities are indeed some form of energy, requiring them to be studied in hopes of detecting the actual nature and frequency level of that energy.

Hans Holzer, author of *The Supernatural*, believes that ghosts are trapped in a state between two worlds. One of those worlds is ours, and the other is a "world next door," which suggests either a universe parallel to ours, or another dimension of space and time. Ghosts remain in this in-between-land because, Holzer states, they have unfinished business. Perhaps the trauma or emotional turmoil involving their death creates a shock to the system that does not allow them to "cross over," as a popular medium named John Edwards likes to call it.

These after-death communications, or ADCs as author Peter Novak (*The Lost Secret of Death*) puts it, have been investigated and catalogued in a massive private research project by Bill and Judy Guggenheim. Analyzing thousands of ADC reports from around the world, the Guggenheims created a literal database of departed souls who come back for a brief time to give a message to a living loved one. These ADCs differ from the many reported cases of religious figures appearing as whole entities, such as the many sightings of Christ or the Virgin Mary. It is suggested that these apparitions represent a different type of phenomena, or, in the minds of many skeptics, a different type of mental gymnastic or mass hysteria.

Novak equates "haunting ghosts" with his theory of the binary soul doctrine, which suggests the two halves of the soul are split. The disembodied unconscious mind loses its conscious half and remains trapped in a "fixed dream world formed out of its memories and emotions," without any objective awareness, analytical reason, or even the ability to communicate verbally with the living. In *The Lost Secret of Death*, Novak suggests ghosts are the polar opposite of near-death experiences, or NDEs. Ghosts are immersed in their memories in our world, while near-death experiencers find themselves immersed in a world where they are liberated from memories and emotions. "Each seems to be the missing half of the other."

But not every violent or traumatic death results in a ghost, and not every ghost died a violent or traumatic death. Many sightings involve auditory phenomena, such as knocks and bangs around an old house, breaking glass, or the repeated sound of music played on some far-off instrument. Usually, a witness reports hearing footsteps and/or voices of people who are not in the room, or, in the case of my nephew when he was a child, hearing voices in the walls of his bedroom. (The jury's still out on whether he was having a ghostly visitation or a bad dream!)

Holzer notes one such occurrence at the residence of a fine colonial house in Stamford, Connecticut, built during the American Revolution. The current owners had lived in the house nearly 10 years and often heard, usually late at night, the sounds of a full orchestra, "as though a radio were on in another part of the house." Obviously, after checking to make sure no radios had been left on, the owners (who also reported loud knocks and crashing sounds in various parts of the house and who had visual ghost sightings as well), called Holzer in to investigate.

EVP and Other Ghostly Forms

Are these sounds also a type of energy trapped between two worlds? EVP, or Electronic Voice Phenomena, is a process that captures the voices or sounds of the dead onto magnetic recording tape. EVP is not new to parapsychologists, many of whom, like the International Ghost Hunters Society, have been investigating the phenomenon for years. The embedded "ghost" voice can be heard when the tape recorder is played. Often the voice is clear as a bell, although sometimes it is lost in the range of background noise. EVP investigators use the highest quality digital recorders when they can, but some EVP cases have occurred on cheap tape decks purchased at a local Target or Wal-Mart. Only with the release of the movie *White Noise*, starring Michael Keaton and based on the book *Miracles in the Storm* by Mark Macy, did EVP enter the mainstream public awareness, no doubt resulting in a slew of new organizations, investigative clubs, and Web sites devoted to posting EVP information, and even actual recordings.

And lest you think ghosts take on the form of only people, animals, or plasma orbs of energy, there are reports (albeit rare ones), of ghostly cars and phantom trains. Usually spotted near the scenes of tragic accidents, "auto-apparitions," as I call them, were preceded by numerous sightings of horses and carriages. Now that we drive around in four-wheeled vehicles instead of four-footed ones, there are occasional reports of cars that appear and reappear on roadways, on bridges, or in fields, and trains that roar into and out of visibility. One such sighting occurred in 1879 at the site of a Scottish rail disaster. A train plunged into the river below the Tay Rail Bridge, killing all on board. Since then, numerous witnesses claim that they've see a spectral train travel halfway across the bridge, and then disappear.

Ghostly planes are commonly sighted in Europe, usually vintage war fighters, and many a seaworthy witness has come in contact with a ghost ship. The *Flying Dutchman* of the 18th century comes to mind.

The Science of Ghosts

Perhaps the energy trapped in ghostly form does not have to be bio-physical in origin, or perhaps these haunted vessels and vehicles serve as containers for the energy of the victims who lost their lives. In any case, the term *ghost* is not always indicative of a dead person or pet trying to reach out from beyond. Some native traditions, such as the Banks Islanders of the Pacific, believe that objects similar to stones and rocks can harbor ghosts.

Whatever ghosts are, and wherever they come from, serious investigators realize that the only way to give credibility to the phenomenon is to take a scientific approach. The presence of energy requires an under-standing of electromagnetic fields, or EMFs, which can indicate fluctuations in the energy field of a certain location, such as a house or graveyard. Many investigators believe that people may experience paranormal phe-nomena resulting from their exposure to EMFs, suggesting that perhaps the witnesses themselves may be a part of the trigger for ghostly apparitions. This theory claims that EMF effects on the human brain may actually evoke experiences in "sensitives," people who seem to attract paranormal events, and often exhibit ESP and other mental abilities beyond the norm.

Ghost hunters such as Joshua P. Warren, author of *How to Hunt Ghosts* and *Pet Ghosts: Animal Encounters From Beyond the Grave*, and president of LEMUR (League of Energy Materialization and Unexplained phenomena Research), treat the subject as any other scientific field. The goal, according to Joshua, is to accumulate well-documented cases and create a database of hard evidence. Additionally, the research team is attempting to create ghostly phenomena in a lab setting.

Joshua divides ghosts into five categories: entities, imprints, warps, poltergeists, and naturals. Joshua told me that, just as some aspects of humans exist in another dimension (one of mind), ghosts seem to primarily occupy some other dimension as well. "In fact," he stated, "they may

Left to right: Joshua P. Warren and LEMUR members Casey Fox and Brian Irish investigate a haunted house. A mysterious orb appeared next to Fox's head when meters detected anomalous surges of electromagnetism. Courtesy of Forrest Connor.

primarily reside in the same dimension as that of our minds. Understanding how and why they manifest relies on both understanding that realm, and determining what conditions in our physical realm bring those two dimensions temporarily closer." He sums it up by saying that ghosts are "non-physical entities" if we define non-physical as "not being restricted to the known laws of physical matter." By using the scientific method and creating a collective database, Joshua believes we may one day have enough data to isolate the patterns and correlations that will finally realize the "essential conditions for spectral interactions to occur."

Serious ghost hunters such as the team at LEMUR use specific techniques and types of equipment in their quest to obtain solid evidence of ghostly sights, sounds, and frequencies. Just visiting haunted tourist traps provides little in the way of scientific proof for those who truly seek to understand, not just experience, ghostly phenomenon.

Tools of the Trade

Capturing a spectral presence on film or tape recorder is the goal of ghost hunters who seek out haunted places. Serious investigators use the following equipment to try to catch an apparition in the act:

- ID (comes in handy if the police show up, and please don't trespass).
- Flashlight and batteries (to find your way in the dark!).
- Watch (to record times of events).
- Compass (to find your way in a big area).
- Still camera with 400 and 800 speed film (black and white is best).
- Infrared film (for more experienced photographers).
- Video camera (for recording movements and layout of locale).
- Notebook and pens/log for recording times and events.
- Electromagnetic Field Detector (most important piece of equipment—for measuring fluctuations in the energy field).
- Digital tape recorder (for recording background noise and/or taking notes).

Tape recorder with external microphone and extra tapes.

Infrared meter (for displaying levels of infrared activity).

Thermometer/thermal scanner (to detect temperature rises/drops).

Night vision scope (ghosts like to come out and play at night).

Candles and lighter (in case the flashlights don't work).

Hand-held radios (for communicating with fellow investigators).

Motion detectors (to detect movement other than your own!).

Cell phone, snacks, and first-aid kit (for the investigators, not the ghosts!).

A friend or two (ghost hunting is better with a team). All the best equipment in the world won't find you a ghost, though, unless you know how to use the stuff.

Once an investigation is conducted and added to a growing database, researchers can look for patterns to emerge, indicating what means ghosts are using for moving between our world and theirs. Theories abound, just as in the field of ufology, but if enough evidence points to, say, interdimensional travel or the role of human brain activity in manifesting physical events, that is where more time, energy, and resources can be placed by those involved in the pursuit of the paranormal.

Warren and LEMUR emphasize the need for scientists to conduct personal research, and they show how that is easier than ever thanks to the enhanced communication of the Internet. But he, and countless other researchers, see the scientific community shunning the paranormal as "fringe science," prompting Joshua to suggest that funding might be part of the problem facing any scientist who does want to confront the phenomenon in an intelligent and open-minded manner.

"The next problem is overcoming the political resistance to development of those institutions upon which scientists rely for funding." Scientific egotism is a problem Joshua sees as the obstacle between progress and retaining the status quo. "Scientists who don't understand the paranormal don't understand science," he explains, "Miracles such as turning water to wine or resurrecting the dead, are paranormal. And experiments proving time travel, teleportation effects, and placebo results, are also paranormal."

If and when the scientific community openly embraces the possibility of ghosts, guys such as Joshua P. Warren will be waiting with massive databases of collected case studies to offer up for study. In the meantime, he'll continue to do what he's always done: investigate each and every case that he can with passion, excellence, and a dedication to the scientific method that may one day reveal the truth. "With our current technology, we cannot expect to solve this mystery here," he states in *How to Hunt Ghosts*, urging those coming into the field to apply what they learn to "new, original, and innovative means of breaching the gap between physical and spiritual reality."

Perhaps we are on the verge of achieving that breach. New discoveries in the world of quantum physics excite Joshua, especially as they pertain to his quest for understanding what ghosts may be, and where they may be coming from. We'll hear why in following chapters.

Poltergeists

Ghosts aren't the only form of spectral energy reported all over the globe. Poltergeists (German for "noisy ghosts") differ from the traditional ghost in that they often occur around one person, usually an adolescent female or teenager. Researchers believe that somehow these persons, also known as "agents," manifest telekinetic bursts of energy in the form of moving objects or manipulations of their physical environment, suggesting that *the source of the energy is the agents themselves.*

Parapsychologists once held poltergeist activity to a certain level of standards in order to label it as such:

The activity is usually centered on one person.
The activity is brief in duration and non-sustaining.
The activity disturbs the physical environment around that person.
The activity usually ends and does not follow the person throughout life.

But not all experts agree. Some, such as Hans Holzer, feel that poltergeists are just "the part of a haunting involving noise or physical movements." Others, such as Joshua P. Warren, define a poltergeist as the "repeated manifestation of ghostly activity" that is primarily dependent on one or more specific persons. In my experience with my friend's poltergeist, there was also my mysterious "chill" with the cold energy moving across me, which could be defined as a ghost, rather than a burst of energy from the person associated with the events, who was not even present that morning.

In fact, the presence of both poltergeist and ghost activity surrounding the same person/locale is not all that uncommon, according to numerous books on the subject, including Janet and Colin Bord's *Unexplained Mysteries of the 20th Century*. The Bords document a number of cases involving mischievous spirits terrorizing families from France to South Africa to America. Banging noises, flying dishes, doors and windows locked and unlocked, vibrating partitions, even flying horse collars have been reported, along with some standard ghostly apparitions, in homes across the globe. One involved a farmer and his family in 1916 who were driven from their home in Ireland by objects crashing against the walls and floating near the ceiling. Interestingly enough, the farmer who reported the activity, and abandoned the farm, had just experienced the loss of a child, indicating that perhaps the dead child's energy was responsible, trapped within the home's confines. Or perhaps it was centered on one of the farmer's other surviving children. In any event, the activity ceased once they moved to a new location some distance away.

Another case involved a woman and her foster child, who began experiencing an outbreak of smashed dishes and moving furniture in 1976. Here, a 9-year-old child was present, but other cases involve families that move into homes where the activity seems to emanate from the home itself, as if the home is the "agent." This is what happened in 1948 in Vachendorf, Germany, when a refugee family lodging in an old mansion was bombarded with stones, tools, and other objects. Perhaps this old mansion was indeed "haunted" and needed the trigger of a family to set the noisy spirits free . . . or perhaps the family itself was the agent, and the house was just obliging. Which came first, the chicken or the egg?

Most, if not all, poltergeist activity suggests a distinct connection between the brain activity of a human being and the outer environment (usually in the form of telekinesis). This was documented in numerous cases, including the events surrounding a woman named Tina Resch, resulting in the book, *Unleashed: Of Poltergeists and Murder: The Curious Story of Tina Resch* by William Roll, Ph.D. (with Valerie Storey). Roll himself witnessed the phenomena, and spent several weeks with the Resch family documenting the strange events surrounding the 14-year-old girl.

The activity would last for several years, culminating in a tragedy that would leave Tina behind bars. But Roll became convinced that the borderland world between mind and matter was something that serious investigators could replicate in a lab setting. Roll found connections between Tina's telekinetic abilities and higher-than-average geomagnetic disturbances. Because the human body literally exists in a sea of electro-magnetic energy, and the brain itself is highly sensitive to changes in the EM field, Roll suspected this was the reason why many cases of what he termed RSPK, recurrent spontaneous psychokinesis, occurred in con-junction with higher solar activity, such as magnetic storms and flares.

Roll was drawn to a theory being discussed on the cutting edges of physics as a potential "fuel" for the energy displayed by Tina: Zero Point Energy. Roll felt this "infinite sea of electromagnetic energy" that filled space could also be used, albeit inadvertently, by people such as Tina who displayed PK ability. Zero Point Energy was also being explored by Hal Puthoff as a future source of energy for space travel.

Many poltergeist "agents" also display ESP abilities and other talents associated with the mind, which we will take a look at in Chapter 4.

While scientists continue to relegate ghosts and poltergeists to the fringes of pseudo-science (most likely due to their own inability to explain it), one area of the paranormal baffled even the most experi-enced investigators with the most open of minds. Ghosts and poltergeists are only one side of a coin that, when flipped, leads to a bizarre study known as cryptozoology.

Cryptozoology

Everyone has heard of Bigfoot and the Loch Ness Monster, and even the legends of Mothman and the Jersey Devil have made it to the big and small screen. But along the strange highways and byways of life there exists a parallel universe right here on earth, filled with weird and inexplicable creatures that, more often than not, materialize out of thin air, and vanish right back into it.

We're not interested in those sightings that involve solid, physical animals or beasts that could very well be living in the shadows of the developed world. Consider the mighty Coelacanth, the "living fossil" thought to be extinct more than 65 million years ago. One was found alive and kicking in 1938 at the mouth of the Chalumna River on the east coast of South Africa, suggesting that the "fish out of time" had survived by staying in the deep, dark waters of the Indian Ocean, undetected until that one fateful day.

Perhaps, similar to the Coelacanth, creatures such as Sasquatch, Nessie, and the giant flying birds with 15-foot wingspans reported in the 1970s from Alaska to Puerto Rico are simply "beasts out of time," undetected by the public en masse, until we either develop their swath of forest or dam up their lake. The big cats that terrorized the British Isles most likely fall into this category as well, as perhaps do the "BigHoots," giant owls reported in parts of England and America during the 1980s. This is not hard to believe, especially in light of the many discoveries including amazing oversized insects and entirely new species of flora and fauna as we literally flatten the rainforests. Who knows what could be hiding in there?

Some of these creatures, according to witnesses, have the characteristics of ghosts. These phantom dogs, spectral cats, and half-man/half-beasts have been reported for centuries in dozens of cultures and countries, defying scientists and paranormal experts alike. It is one thing to accept the possible trapped energy of a dead person who didn't want to die, but quite another to wrap your mind around a red-eyed beast that appears in your hallway and vanishes in a burst of sparks.

The Mothman legend comes to mind. Documented in dozens of books and even a spooky movie with Richard Gere and Laura Linney titled *The Mothman Prophecies*, which is based upon John A. Keel's book of the same name, sightings and encounters with a giant gray man with wings and red glowing eyes surfaced in 1966 in Point Pleasant, West Virginia. Coinciding with numerous UFO reports, and even allegations of poltergeist activity, the Mothman appearances terrified a number of witnesses who reported the creature at various locales near the Ohio River Valley.

The eyes were described as "hypnotic" and the creature was often described as having the ability to fly at high speeds. After numerous sightings throughout West Virginia, a reporter dubbed the term "Mothman" and more reports poured in, many of which are documented in Keel's book, and in the book *Jerome Clark's Unexplained!*.

Other witnesses reported UFOs and electrical interferences at the same time the Mothman was spotted in the vicinity, and even a report of a dead dog that may have been a victim of the creature was noted. Eventually more than 100 witnesses would talk to John Keel, with reports tapering off during 1967. Sightings of the huge flying man with bat-like wings and eyes trickled in through the following decade.

Keel, by the way, would in his many years as a paranormal investigator and UFO researcher, dismiss the concept of the standard extraterrestrial theory, suggesting instead that UFOs, aliens, demons, angels, and other paranormal entities and events all involved a "complicated system of new physics related to theories of the space/time continuum." Keel also believed, as did Jacques Vallee, that paranormal objects and apparitions may not even be constructions of matter, but instead may involve the consciousness and perception of the witness.

This may also explain sightings of a strange creature called the Chupacabra. The name means "the goat sucker" in Spanish and refers to the creature's preference for goat's blood. The Chupacabra, now the stuff of legend, is described as a bipedal creature anywhere between 3 to 6 feet tall, with a lizard-like tongue, spines down its back, and red glowing eyes. Some reports claim the creature has wings, but all seem to agree this is one mean monster. Elusive as the Bigfoot and as frightening as the Mothman, the Chupacabra attacks animals in local villages in the southern

Paranormal researcher/author Nick Redfern mugs it up with a baby Chupa. Courtesy of Nick Redfern.

part of the United States, Mexico, and Central America, and sucks their blood. The attacks first began in Puerto Rico, but soon spread throughout Mexico, Central America, South America, and even into the southern United States. The wildest reports came out of Chilé in the year 2000, when locals claimed several Chupacabra were actually caught and carted off to some top-secret government facility.

There are many theories about the Chupacabra: Some say it is the discovery of a new species; others think there is a portal to another dimension through which these creatures move between worlds; and a third possibility that they are actually aliens deposited on our planet. The reports of glowing eyes suggest a possible link to creatures such as the Mothman and spectral black dogs and giant cats with red glowing eyes that have been reported in various countries. These creatures take on a physical appearance, although often they are reported to "flicker in and out of view" or take on a see-through transparency after awhile. Similar to ghosts, they appear to be trapped energy from somewhere else, trickling into our world from some unknown source or entry point. And lest you think it is only dogs and cats, there are even reports of spectral horses (the White Devil is said to haunt the United States deserts), sheep (a ghostly ewe is said to haunt one of the oldest houses in Orkney), and headless pigs.

The British Isles are notorious for a plethora of black dog legends. According to local lore, legends go by various names such as Black Shuck, Hellhound, Gurt Dog, Black Angus, and CuSith, which is Scottish for "faery dog," There have been reports of these creatures from as far back as the 17th century, and according to a fascinating book by Nick Redfern, they are still haunting villages and moors, and scaring locals out of their wits.

The book *Three Men Seeking Monsters* follows Redfern and professional monster hunters Jon Downes and Richard Freeman on a six-week trek to seek out the "werewolves, lake monsters, giant cats, ghostly devil dogs, and ape-men" that plague the British countryside. Their journey brings them face to face with dozens of eyewitnesses and hundreds of written reports concerning encounters with phantom animals and spectral beasts with glowing red eyes (some even allegedly half-human), such as the spectral cats that roam the Rendlesham Forest. Rendelsham is the sight of one of the most famous UFO landings and cover-ups in history (documented in Jenny Randles' books *Sky Crash—A Cosmic Conspiracy* and *From Out of the Blue*). Then there is the black dog described by witnesses in Woodbridge, England—"They were even more shocked, however, when a moment later it reappeared and proceeded to 'flicker on and off' four or five times before vanishing permanently," this accompanied by the smell of burning metal.

As Redfern, Downes, and Freeman get closer to the phenomena, they encounter an intriguing witch named Mother Sarah, who informs the men that what they are dealing with may not be "of this earth" at all, but from somewhere else entirely. In fact, Mother Sarah seems to have an inside scoop on the relationship between matter and consciousness, suggesting that our belief and perception could bring about the manifestation of these entities.

Mother Sarah's story, based upon a book supposedly passed down through generations of her family, contains bizarre elements of humans controlled by emotional parasites and manipulated by a phenomenon called the Cormons. These strange creatures would be manifested and utilized in battle against opposing forces, but somehow over time, these Cormons stopped obeying marching orders so to speak, and began to run rampant over Great Britain, appearing and disappearing "free of whatever restrictions had tied them to their previous plane of existence."

As odd and out there as Mother Sarah's story might sound (and do read *Three Men Seeking Monsters* to get the entire lowdown), it touches upon beliefs shared by Vallee and even Jung that the collective unconscious can work in tandem with physical energy to create physical phenomena that acts and behaves "paranormally."

Mother Sarah tells Redfern and colleagues that "the ability of the creatures to exist in our environment was directly linked with our ability to perceive them." She touches upon the most cutting-edge theories of physics when she suggests that these entities are from a realm far closer than extraterrestrial space.

Her comments about emotional parasites reminded me of Colin Wilson's horrific story of psychological terror, *Mind Parasites*. If we humans are nothing but host bodies for entities that feed off of our consciousness and belief, that would explain a lot about history's darker trends, such as the lust for war and fear-based paradigms. It would also explain why sightings of UFOs and black dogs seemed to occur in waves and cycles, driven by the mass consciousness of humans, in a sort of contagion.

Akin to the tipping point of social, political, and economic trends, perhaps there is a tipping point for paranormal phenomena.

Strangely enough, quantum physics and the new science of consciousness might suggest how and why.

But before we ride off on that broomstick, I have two tickets for a cruise to the Bermuda Triangle.

Care to join me?

Into the Void:
Energy Vortices and Zones of High Strangeness

Sit down before fact as a little child, be prepared to give up every preconceived notion, follow humbly wherever and to whatever abysses nature leads, or you shall learn nothing.
—Thomas Henry Huxley

Do Not Enter.
—sign on an average teenager's door

There are places on Earth where the laws of physics and the natural order of things just don't seem to apply—places where planes and ships regularly disappear off the face of the earth and are never found again; places where people repeatedly see mysterious creatures in the woods, or flying objects in the sky; places where time bends and space distorts.

The Bermuda Triangle

The best known of these places of high strangeness is the Bermuda Triangle, a zone of mystery that stretches from the tip of Florida's southernmost point, then north to the island of Bermuda and south of Puerto Rico. This huge swath of Atlantic Ocean is noted for a high incidence of unexplained disappearances of ships, small watercraft, private planes, and military aircraft, the majority of which have never been recovered.

The two most widely known disappearances in this area are the *Cyclops* and Flight 19. The USS *Cyclops* was a collier launched on May 7, 1910, and placed into service in November of that same year. During the first World War, she was assigned to the Naval Overseas Transportation Service, and later helped to fuel British ships in the South Atlantic. En route to Rio de Janeiro, Brazil, in 1918, she stopped at Barbados on March 3. She departed on March 4 and vanished without a trace, along with 306 crew members and passengers, somewhere in the Triangle.

Some 17 years later, the most legendary of Triangle tales would occur on the afternoon of December 5, 1945, when five TBM Avenger Torpedo Bombers would depart from the U.S. Naval Air Station in Fort Lauderdale, Florida, at around 2:10 PM on a navigational training flight, and go down in the history of mysteries when all five planes disappeared out of the skies, never to be found again.

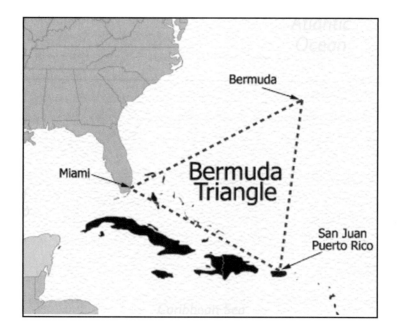

Flight 19 would make radio contact two hours later, indicating the pilots were experiencing compass malfunctions and that the instructor, Lt. Charles Taylor, believed himself to be somewhere off the Florida Keys. He and his student pilots, 13 in all, would be urged by Lt. Robert Cox, another pilot flying in the vicinity who picked up the radio transmission, to fly north toward Miami, but the more Taylor tried to fly north, the farther out to sea the Avengers actually moved. A radio transmission picked up on the mainland indicated that one of Taylor's students, a superior officer named Captain Edward J. Powers, suggested flying west, but the Avengers moved farther north and east, as if Taylor could not get his visual bearings.

By 5:50 PM, the ComGulf Sea Frontier Evaluation Center got a fix on Flight 19's radio signals, showing them east of New Smyrna Beach, Florida, but communications were so poor that the instructions to redirect the planes could not get through.

By 6:20 PM, several craft were sent out after Flight 19, including a Dumbo Flying Boat and two Martin Mariners, one of which never arrived at the search area (although crew members of the SS *Gaines Mill* reported seeing an explosion and finding airplane debris in the same general location). The final radio transmission from Flight 19 came in at 7:04 PM. The aircraft was never heard from again.

The five planes and any related debris have never been found, and the bodies of the 13 pilots were never recovered, despite extensive, ongoing searches that continue to this day, including several deep-sea dives, which were filmed as documentaries and aired on several cable channels, including the Sci Fi Channel, Discovery Channel, and Travel Channel.

The Bermuda Triangle got its name from writer Vincent Gaddis in a 1964 issue of the fiction magazine, *Argosy*. But it was Charles Berlitz's 1974 best-selling book, *The Bermuda Triangle*, that created a buzz (due to its highly sensationalistic nature), selling millions of copies and leading to dozens more books, movies, and documentaries, not to mention Web sites devoted to either perpetuating or debunking the enigma. In 2006, the Sci Fi Channel aired a mega-million-dollar mini-series called *The Triangle*, along with documentaries featuring new evidence, theories, and even interviews with the likes of theoretical physicist Michio Kaku.

Although skeptics and scientists alike have written off the Bermuda Triangle as a place of nothing more than "extreme weather," "human error," and "choppy seas," the disappearances continue, often defying any rational explanation. In 1975, the publication of a book by Larry Kusche titled *The Bermuda Triangle Mystery—Solved* attempted to put the entire enigma to bed as nothing more than a series of misinterpreted data; the author did succeed at explaining that many of the strange disappearances were not so strange at all, rather the result of poor research and lack of facts. But the enigma continued, and dozens more stories emerged that defied Kusche's claims that the myth was forever debunked. If Berlitz could be accused, as some did, of making a mountain out of a molehill, then Kusche and other debunkers could be accused of making a molehill out of a mountain. Neither side had all the answers.

Theories abound of extraterrestrials swooping up planes and ships, huge fog banks swallowing up everything in their path, portholes and stargates to other worlds, and even a warp in the fabric of space/time. As crazy as those theories may sound, modern science may actually lend some credence to them, but we will get to that later.

There are some interesting facts to keep in mind about the Triangle that suggest more acceptable explanations. The deepest point in the Atlantic, the 30,100-foot Puerto Rico Trench, is located within its boundaries, which could account for why many vessels and craft are never recovered.

And the Triangle is one of only two places on Earth where a compass actually points to true north (normally they point to magnetic north), which could account for much human confusion and error, especially among pilots who easily lose their visual bearings amidst stormy conditions, or when out to sea far enough where land is simply not discernable. The other such place that shares the magnetic characteristics of the Triangle is known as "Devil's Sea" off the east coast of Japan, and is also the site of many mysterious disappearances.

The Naval Historical Center's Web site claims many of the disappearances can be credited to environmental factors, including the presence of the Gulf Stream, which can be so swift and turbulent it can wipe out evidence of a downed craft. "The unpredictable Caribbean-Atlantic weather pattern also plays its role," the NHC states, even citing the topography of the ocean floor, which varies from lengthy shoals to deep marine trenches, as potential navigational hazards.

The Science of the Bermuda Triangle

One of the most interesting new "scientific" attempts to prove the Triangle is not so mysterious comes from British geologist Ben Clennell, who believes bubbles are to blame. Clennell's theory involves giant gas bubbles released from underwater landslides. This frozen methane gas, also called "gas hydrate," is produced by deep-sea bacteria that exists on the ocean bed and, when released to the surface, could be disruptive enough to take down a big ship. The "gas burp" theory (as I like to call it) has been backed up substantially with laboratory experiments done at Monash University in Australia and has been reported in the September 2003 issue of the *American Journal of Physics*. More evidence lies in the discovery of actual vessels proven to have paid a visit to Davey Jones' Locker thanks to giant gas bubbles in other parts of the world. In 2000, at a place called the "Witch's Hole" in the North Sea, a documentary film crew discovered a 22-meter steel-hulled fishing trawler sitting in the middle of an underwater crater. The crater was determined to have been the cause of a gas burp, which would explain why the ship was found in almost undamaged condition, sitting horizontally, as if it just plain sunk.

Even the U.S. Geological Survey got into the gas burp act, publishing a paper in 1981 about the appearance of gas hydrates off the southeastern United States coast. Other researchers would speculate that methane gas might also interfere with the functions of altimeters in aircraft, interfere with the lift of air needed to keep planes flying, and potentially disrupt the mix of fuel and air, causing combustion to stop and engines to stall.

For every scientific or "rationale" theory explaining the Triangle, there are events and disappearances that defy all rational explanation. Author and researcher Gian J. Quasar has devoted over a decade of his life to meticulously investigating the phenomenon of the Triangle. His book *Into the Bermuda Triangle* dissects case study after case study of over 1,000 ships and aircraft that have vanished in the Triangle over the past 25 years.

Quasar meticulously examined official reports and records from the National Transportation Safety Board (NTSB), Coast Guard, and various government agencies. He also interviewed scientists and actual Triangle survivors, putting together what is no doubt the most comprehensive study of the Triangle to date. In addition to the *Cyclops* and Flight 19 cases, he closely examines dozens of other mysterious disappearances from the days of Columbus in 1492, whose ship logs reveal his own close encounter in the Triangle, to modern military aircraft and luxury cruise ships that never made it back to port.

Some of the cases are easily solved when searches reveal the telltale crash signs of debris and oil slicks. Many of the cases involve EMF interference, UFO sightings, and even the presence of a thick fog that appears out of nowhere. The most enigmatic case of all is one Quasar found where a "121.5 EPIRB (Emergency Position Indicating Radiobeacon)" was picked up on May 4, 2002, just 50 miles south of Bermuda. A quick overflight by other craft found no trace of the plane, even though the signal continued to be received, as if, the author states "from another point in space and time."

From Cessnas to Ventura military craft to DC-3s and F-4 Phantoms, from tall mast ships to tankers to military warships to cruise ships, the sheer number of craft lost to the Triangle is astounding. Skeptics claim this area has a high loss rate simply because it has a high travel rate, but Quasar and other researchers point to the numerous other high traffic spots on the globe that lack the number of vanished craft. That most of these planes and ships are never recovered, nor their crew ever found, suggests that the deep waters of the Triangle might hold more mysteries than the most skeptical of skeptics is able or willing to admit. And when there are actual survivors to tell their experiences, it becomes even harder to pass them all off as human error in high waters.

Don Welch and Ron Reyes are two of those survivors. In the book *Into the Bermuda Triangle*, they share their story of setting out from Port Canaveral, Florida, in 1996, flying into moderate winds and seas at 2–3 feet. Approximately 5 miles offshore, they entered a small fog bank that enveloped them momentarily, releasing them into an area of clear skies and calm winds, with the fog around them almost like the eye of a hurricane. "We went from a moderate wind into a fog bank and then into an opening in the fog bank which was a dead calm with sunshine and no sounds." After then proceeding back through the fog, they again encountered the moderate winds. Both men agreed the situation left them feeling "eerie."

Encounters and Disappearances

Other creepy encounters with strange fog banks include the report from retired Coast Guardsman Frank Flynn, who was aboard the *Yamacraw* cutter in 1956 when he and the crew encountered a vapor mass so dense, Flynn claimed that the beam of their search light reflected off of it. After a few attempts to "nudge" the mass with the ship's starboard side, they entered the mass, at which point their engine room reported losing steam pressure. They reversed course and departed the mass, and their engine and electronic gear returned to normal.

Even Charles Lindbergh, the great aviator, reported an unusual haze and magnetic anomalies in 1928 while piloting his famous plane *Spirit of St. Louis* in the Triangle. He wrote about the incident in his 1976 book, *Autobiography of Values.*

But perhaps the most chilling encounter happened to father and son real estate brokers who had an amazing experience in December 1970. Bruce Gernon, Jr. was flying with his father and another business associate. They took off from Andros Airport in the Bahamas, flying a Beechcraft Bonanza A36. Shortly after takeoff, Bruce noticed an elliptical cloud in front of their craft, about 1 mile away, hovering at about 500 feet above ocean level. "It was a typical lenticular cloud," Gernon told Quasar in an interview for Bermuda-Triangle.org, a Web-based research site. The weather was good, reported the Miami Flight Service, but the lenticular cloud changed shape before their eyes into a huge cumulus cloud, which seemed to be rising at the same rate as the plane.

The cloud engulfed the aircraft and after 10 minutes of climbing in and out of the cloud, Gernon, who was piloting, entered clear skies at an altitude of 11,500 feet. Cruising at approximately 195 miles per hour, he looked back and noticed the cloud had formed a "giant semicircle extending around us." He estimates the cloud's diameter must have been over 20 miles long. They noticed another cloud building up in front of the craft that looked like the one they had just left, only with a top height of at least 60,000 feet, and upon entering that cloud, it became "dark and black, without rain . . ." They described seeing extraordinarily bright flashes that lit up the area and revealed that the curve of the cloud continued all around them, similar to a tunnel, and that the cloud they left behind was really only one end of the ring-shaped mass they were now inside!

It took another 13 miles of travel before Gernon and passengers saw a u-shaped opening, which closed into a hole as they approached it. They exited the "tunnel" after about 20 seconds, during which Gernon experienced a sensation of weightlessness and the total malfunctioning of their electronic and magnetic instrumentation. Gernon was able to make contact with a radar controller in Miami, and he reported they were about 45 miles southeast of Bimini, but the controller couldn't identify them in that area.

Gernon described it as "entering an electronic fog," where the air outside took on a dull, gray-white haze and no sign of the ocean, horizon, or sky was visible. He approximated they were in this fog for 3 minutes before the radar controller radioed back that he had an aircraft directly over Miami Beach moving west. But Gernon's watch told him only 34 minutes had passed since take-off. They simply could not be that close to Miami Beach, but when the fog suddenly broke in a manner he described as "long horizontal lines . . . widened into slits about 4 or 5 miles long . . . continued expanding and joining together," until all the strange slits had joined and the gray haze disappeared, replaced by blue skies and the barrier island of Miami Beach below.

Somehow, the Beechcraft had flown about 250 miles in only 47 minutes. Something wasn't right, and Gernon says he and his fellow passengers didn't talk about why for a long time.

Gernon also reported a phenomenon well known in the area called the "Green Flash," and even admits to having UFO sightings heading out towards Bimini. His story is one of many on record of strange fog walls and clouds that envelop air and sea craft and wreak havoc on electronic instrumentation and magnetic measurements. The theories behind these possible "time storms" and "electronic fogs" include possible portals to another dimension where UFOs travel freely, and a connection with the ancient city of Atlantis. The discovery of huge, manmade rocks known as the Bimini Road added fuel to the fire, suggesting that the fogs are caused by energies being released by Atlantean power generators deep beneath the sea.

But the Bermuda Triangle and its sister site, the Devil's Sea, are not the only known hot spots of paranormal mystery on the planet. Folks living near the Great Lakes in the United States have their own zone of lore and legend.

Ships have been disappearing in the Great Lakes under bizarre circumstances since 1889, when stories arose of tall, masted ships that set sail and never arrived at their destinations. Many of these disappearances occurred in Lake Ontario, which is said to be home of the Marysburgh Vortex, a strip of water at the lake's eastern boundary that, according to some researchers, swallows up a higher concentration of ships than the Bermuda Triangle. In 1889, the disappearance of the tall-masted ship *Bavaria* set off a search into the lake's eastern waters. The captain and crew of the *Armenia* would find the *Bavaria* sitting on a small shoal, but, when they boarded, they found no evidence of the crew except a batch of freshly baked bread in the galley and an unfinished repair job on deck, indicating that whatever happened to the crew, happened quickly.

This spawned a history of fated vessels in the region, such as the *Quinlan*, which actually smashed into the Marysburgh shore after a violent storm. The few survivors spoke about strange anomalies such as compass malfunctions that made navigating the raging waters impossible. In the 1960s and 1970s, experienced pilots reported encountering strange fogs that blocked out all visuals of land. Tom Walker, a veteran flyer who took off for a simple flight to Owen Sound, 80 miles north of Lake Ontario, ending up in a local hospital, unable to account for anything that happened once his plane entered a cloud of fog.

Vile Vortices

According to researchers for the local Mutual UFO Network, and a book titled *Canvas & Steam on QuinteWaters*, two-thirds of all ships lost in this area go down in the eastern end of Lake Ontario. The U.S. Navy and Canadian National Research Council began a study in 1950 of anomalies in the region, resulting in an investigation by Wilbert B. Smith and a team of scientists working under the Canadian Department of Transport. Their research led to the discovery of an area with "reduced binding" in the atmosphere near the eastern shore, which may have been responsible for peculiarities in gravity and magnetism within these somewhat mobile "columns" of weird air. These columns were funnels measuring up to 1,000 feet in diameter, from the ocean surface into the sky at high altitudes, as though invisible waterspouts. Inside these columns, which were actually detected by satellite in 2001, magnetic and gravitational data suggested a lessening of the grip of gravity that could move around, yet generally were more common in southern latitudes.

Notice the similarities to the "tunnel" Bruce Gernon entered with his passengers in the Beechcraft.

The question of whether these magnetic vortices are responsible for some of the lost planes and ships led professional biologist Ivan Sanderson to develop his theory of "vile vortices," a series of 12 equally spaced areas around the globe that experienced higher than normal magnetic anomalies and energy aberrations, including strange physical phenomena. These lozenge-shaped areas, which he often refers to as "devil's graveyards," include:

- The Bermuda Triangle
- Japan's Devil's Sea
- Algeria's Megalithic Ruins
- Pakistan's Lower Indus Valley
- Hamakulia Volcano in southeast Hawaii
- Wharton Basin, Indian Ocean
- Easter Island Megaliths
- Great Zimbabwe
- Loyalty Islands, New Caledonia
- Atlantic Ocean off Rio de Janeiro, Brazil
- North and South Poles

These "vile vortices" were described by Sanderson as zones of anomalous behavior, which were set apart at 72-degree intervals latitudinal, where strong currents swirled and hot surface currents from the tropical latitudes met with colder waters of the temperate zone. This produces extreme temperature variables and the potential for high incidences of marine and air disturbances. Sanderson believed that planes and ships could somehow move into a vortex, where a series of natural effects would be triggered, setting off other natural phenomena that might result in electronic and magnetic gymnastics of sorts—the kind that just might down a military jet or sink a cruise ship.

Areas that attract high levels of paranormal activity don't have to be over water. In fact, some of the most concentrated areas for UFO sightings are on land. Reports of cryptozoological creatures and spectral sightings are heavily concentrated in places such as the British Isles, where the moors and forests are filled with things that go bump in the night and the day. The Jersey Devil comes to mind, or the Mothman sightings that plagued Point Pleasant, West Virginia (and were accompanied by UFOs, missing time, electrical disturbances, prophetic visions by locals, cases of strange coincidences and, *Lost* fans, listen up: repeating numbers.

Other virtual vortices for the bizarre include the Bridgewater Triangle in Massachusetts, a 200-square-mile area surrounding the center point of Hockomock Swamp, allegedly called "The Devil's Swamp" by Native Americans. This region alone boasts UFO sightings from 1760 onward, Bigfoot sightings, phantom dogs, cattle mutilations, ghostly activity, spook lights, and even appearances by giant pterodactyl-like Thunderbirds. Remind me never to buy real estate there!

Other Theories

Even well-known places such as the Nazca Lines in Peru, Stonehenge in the United Kingdom, Sedona in Arizona, and Scotland's Loch Ness area may be more than just power spots designated by ancient civilizations or the home to living fossils. The term "geomancy" refers to an ancient form of divination involving soil patterns, but in the 19th century, geomancy was applied to the Chinese practice of feng shui, which at that time was used to determine the location of homes and tombs with regard to topography. Today, we use feng shui, which literally means "wind and water," to balance the energies in our homes and offices, but the concept remains the same: the alignment of "chi," or elemental energy (known in other traditions as spirit, holy breath, or the Force!) with the powerful currents and lines of magnetism that run invisible over the landscape of the earth. Known in China as "dragon current," or lung-mei, these currents of energy could exist in both the positive yang and the negative yin, and, when utilized with purpose, could bring about different desired effects such as harmonious beauty.

This sounds an awful lot like "ley lines," a term coined by local businessman Alfred Watkins in 1921 to explain his theory that ancient British sites had been built in some kind of purposeful alignment. Churches and places such as Stonehenge, the Long Barrows, Standing Stones, and other noted burial mounds all appeared to be constructed on literal lines, many not more than a few miles long, running from one site to another. Watkins believed ley lines might be the remnants of ancient trading routes. Ley lines have since been associated with earth energy lines that intersect places of power and may explain the concentration of bizarre phenomena in certain hot spots across the globe. (Which would also explain why some places have virtually no paranormal events—they just aren't "hooked up!") Many believers insist ley lines follow the cosmic energy patterns of the earth and that "leys" can be detected with dowsing rods.

The time storms and electrical fogs mentioned earlier also occur on land all over the globe. Many of these stunning accounts are featured in a book by noted author and researcher Jenny Randles, *Time Storms: Amazing Evidence for Time Warps, Space Rifts and Time Travel*. The book documents dozens of cases about people entering peculiar energy fogs or clouds and experiencing missing time, potential teleportation, EMF and gravity anomalies, and other paranormal phenomena. These often-glowing clouds suggest a point of entry to a possible "rip" in the space/time continuum right here on earth. The dozens of cases Randles examined all had this one element in common: that a person or vehicle entered a cloud of energy that transported them into another dimension.

Possibly the result of rare atmospheric conditions, such as those suggested by the vile vortices, and with the ability to "warp time," these fogs and clouds could also allow for the manifestation of such elements as ghosts, UFOs, and perhaps maybe even spectral creatures from another place and time. Often the victims return from their time storm experiences in a trance-like state, or, as in the case of Jorge Ramos from Linhares, Brazil, who drove his car into a strange white glow while on the way to a work meeting, they are found in a dreamy state of disorientation five days later and over 600 miles from home! In another case, a man named Roy Wilkinson was walking toward the factory where he worked when he saw an egg-shaped object over a building. The object, milk-white in color, seemed to be surrounded by an opaque haze and changed colors in sequence as he got closer. Roy walked into the haze and felt as if he were in a vacuum; time seemed to slow to a crawl and all sound ceased in what was an otherwise busy area. When the mist vanished, Roy could once again hear the normal sounds and felt a sense of time passing at its usual rate.

These physical symptoms and sensations of being in a vacuum where time stops, sight and sound stop or distort, and consciousness is actually altered, are common elements of time slip and time storm experiences.

Researchers call it the "Oz Factor," after Dorothy's trip to the alternate dimension of the Land of Oz. Many psychologists working with near-death experiences report similar symptoms with patients who have crossed the barrier of death and lived to tell about it.

Canadian brain researcher and specialist Dr. Michael Persinger, who discovered connections between electromagnetic fields and changes to the brain's temporal lobe, established that exposure to EM radiation can induce altered states of awareness and other physical effects such as those described in time storms. The temporal lobe has been linked to out-of-body and mystical experiences, as well as to feelings of dissociation and hallucinations, by neuroscientist Peter Brugger.

The Oz Factor is present in most, if not all, time storm cases, suggesting that the mists and hazes are EM fields generating energy from within the glowing, and often colorful, mass. But it is not just the brain that is being manipulated here. Something is also happening involving the manipulation of linear time. Take the case of two American men who heard a strange noise near the lake they were staying at in Oxford, Maine. They set out at 3 AM to find the source of the noise and found their car enveloped in a colored glow. In the blink of an eye, they were somehow transported a mile away and the car was now pointing in the opposite direction. Now they were surrounded by a grey fog and their car refused to start. Disoriented and lightheaded, they could barely stand up, let alone talk, and both experienced hallucinations in this altered state, lasting for several hours before normalcy returned. Later they would describe their experience to Randles by stating, "We were out of synch with reality."

One of my very first cases as a UFO investigator for my local MUFON chapter was a time storm. I had just passed my field investigator training test and was barely prepared for the high strangeness factor of the report I took from a young man who had lost over two hours of time after entering a thick fog while driving through a local military base to make a delivery. The area, rather remote, does happen to be near the California coast, but this man described the fog as having unusual properties: his watch stopped, his car engine failed, and his sense of direction failed along with it.

Disoriented and unsure of where he was, or even the time of day, he finally came out of the fog in an area miles to the east of where he had been heading. He had somehow lost two hours of time as well, and ended up making his delivery quite later than expected. This young man, whom I felt was sincere and very disturbed by the experience, initially thought he had entered some kind of military test zone, and never mentioned UFOs or abductions, which at the time were hot in the media. In fact, he was rather reluctant to share his story because it did not fit the "normal" pattern of other cases where missing time most usually led to a UFO sighting and possible abduction. Instead, he entered a fog he knew was not normal, even for a coastal native who sees fog on a regular basis.

One of the most frightening cases Randles looked at was actually caught on videotape. Surveillance cameras around a small factory showed a worker approach a gate and disappear into a fuzzy white glow that just appeared out of nowhere. There was a brief electrical disturbance on the camera image, but when they came back to full function, the man was gone. Almost two hours later, the same cameras recorded his sudden reappearance; only he wasn't quite the same man that entered the glow. Now, he was on hands and knees, violently ill, and later would report missing time of two hours.

Clearly these are not incidences of mass hysteria, faked events put on by publicity seekers (most were clearly disturbed by the events), or the result of shoddy reportage. Something was happening in these fogs and hazes that challenges the known laws of science, suggesting that time travel is indeed possible, even if not readily understood. And if these time storms could take on a permanent home, like in the Bermuda Triangle or the Devil's Sea off Japan, that would explain the higher losses of planes and ships, as well as why those few who have experienced the effects of these vortices and come back to speak of them always mention feelings of disorientation, timelessness, being caught in a vacuum, losing sense of direction, hallucinations, nausea, and certainly EMF effects to their engines, compasses, and computer systems.

Many skeptics blow off the whole enigma of energy vortices, lines of power, and fogs that defy space and time as the stuff of science fiction, refusing to look at the number of situations that can't be passed off as the imaginings of the mentally unstable. But just as many "believers" can be accused of turning their backs on facts that might suggest a more "scientific" explanation for these areas of potential "crosstalk" of realities, there just may be openings in the fabric of space and time that lead to other dimensions. These doorways can be where energy moves from one level of manifestation to another, sometimes in chaotic bursts, such as a ghost who flickers in and out of view, and other times in more sustained activity, such as the ongoing disturbances of triangles at sea or glowing clouds that swallow up drivers and deliver them days into the future and miles from where they started.

Windows between worlds.

Because, as we will learn in later chapters, they just might be. Am I blowing your mind yet? Speaking of the mind . . .

Mind, Unlimited:
ESP, Remote Viewing, and Other Mind-Over-Matters

Every man takes the limits of his own field of vision for the limits of the world.
—Arthur Schopenhauer

The human mind will not be confined to any limits.
—Johann Wolfgang von Goethe

There is little doubt that ESP and psychic abilities constitute the most widely accepted category of paranormal phenomena, at least by the public. People of every background and culture believe in the mind's power to reach beyond the five senses and receive information from sources outside the boundaries of space and time. Also referred to as "psi" in the word's truest sense, psychic phenomena have become so widespread and pervasive that serious researchers, including many open-minded scientists, have created a field of study called "parapsychology."

Parapsychology includes the scientific study of events that occur to human beings and animals that include:

ESP Extrasensory perception or "sixth sense"

PK/TK Psychokinesis and telekinesis, which involve mental interactions with matter, often called "mind-over-matter"

Clairvoyance Also known as "remote viewing," the ability to see events occurring at a distance

NDE Near-death experiences

OBE Out-of-body experiences, also called "astral travel"

Precognition Premonitions of future events

Past Life Recall Evidence of reincarnation in memories of past lives

Poltergeists PK phenomena associated with a living "agent"

The Parapsychological Association is an international organization of well-respected scientists from a variety of fields. The PA uses scientific methods for testing psi phenomena to build a comprehensive database of experimental results and to further serious study. On its Web site, the PA states its interest in psi because of its implications:

1. That what science knows about nature is incomplete
2. That the presumed capabilities and limitations of human potential have been underestimated
3. That fundamental assumptions and philosophical beliefs about the separation of mind and body may be incorrect
4. That religious assumptions about the divine nature of "miracles" may have been mistaken

ESP

The most commonly reported psi phenom is ESP. Humans and animals use their five senses to receive information from their environment. Through sight, smell, sound, taste, and touch, we process information and take from it what we need to get through the course of our day. But sometimes, there is a brilliant flash of insight intuition, or the absolute knowledge of a future event, or the persistent memory of a life lived in another time and place, and suddenly we have entered the domain of the sixth sense.

The term "ESP" is said to date back to the late 1880s when French researcher Dr. Paul Joire first used it to describe the ability of a person in a trance or hypnotic state to sense things externally, without the use of the five ordinary senses. Even though ESP and mind abilities exist in writings and religious traditions dating back to the dawn of recorded history, this may have been the first time ESP was used in a more scientific manner. In 1882, the Society for Psychical Research was created in London, England, devoted to the study of members' research, which was published in two society journals.

Soon other countries began their own formal studies, but it wasn't until the 1930s that an American parapsychologist named Joseph Banks Rhine made the term "ESP" popular with the public. J.B. Rhine, a botanist who first used the term "GESP" meaning *general extrasensory perception*, conducted a series of tests using a deck of cards bearing five symbols. These cards were invented by Rhine's research partner, Dr. Karl E. Zener, and eventually became known as the Zener ESP Cards. They consist of a plus sign, circle, square, star, and wavy lines, and were used in a number of different experiments involving subjects either guessing the cards, or having one subject mentally "send" the card symbol to a receiving subject.

The Zener Cards are still a popular method for testing ESP ability.

These card tests were by no means foolproof, and other psychologists and researchers who attempted to duplicate Rhine's tests generally conceded that the experiments were of poor design; but the simple card guessing games did open the door for improvements to experimentation methods, including more modern computer ESP tests, where a subject attempts to outguess the random targets the computer programs.

Other experiments were carried out using sensory deprivation. Called the Ganzfeld Experiments, after a term used in Gestalt psychology to designate the visual field, test subjects were deprived of their sight and hearing, usually with eye covers and earphones. A "sender" would attempt to psychically communicate a randomly selected target and the subject would be asked to match his or her perceptions to the target. The theory behind these experiments suggested that subjects in a meditative or hypnotic state would most likely be able to access psi abilities. A metaanalysis of 25 experiments conducted between 1945 and 1981 on the relationship between hypnosis and psi did indeed turn up evidence that hypnotic induction does facilitate psi performance.

Dream studies also attempt to connect psi ability with altered states of consciousness, such as those conducted at Maimonides Medical Center in New York in the late 1960s and early 1970s. In those experiments, two subjects spent the night in a sleep laboratory. One was designated a sender, one a receiver. The receiver would go to sleep in an isolated room, with brain waves and eye movements monitored. When the receiver entered REM sleep, the experimenter would signal the sender to begin focusing or visualizing a randomly chosen target picture. Toward the end of the REM sleep period, the receiver was awakened and asked to describe his or her dreams, which was then compared to the target picture or image.

Though there was some success in this study project between sender and receiver, other studies designed to replicate the experiment turned up nothing. But the Maimonides experiments suggest that altered states of mind contribute to greater significances of communication between sender and receiver than normal.

Biologist and author Rupert Sheldrake would even carry out extensive studies into the psychic ability of pets, including homing instincts, group communication, and extrasensory perception. He also carried out numerous lab studies experimenting with the ability of plants to communicate with other plants and even with humans. The results were stunning.

There were and still are strong criticisms against this kind of experimentation. Skeptics suggest a possible "file drawer" effect, where only favorable results are published in a given study, as well as the possible influence of the experimenter on the outcome. And of course, there exists the potential for both fraud and shoddy methods of gathering data. The ability to duplicate results as well as finding quality test subjects has eluded many scientists, although as newer and more sophisticated computer experiments are being developed, many of these negative factors could be eliminated.

Another major obstacle is the difficulty researchers experience with obtaining funding for psi studies. Traditional peer-reviewed sources of funding almost never consider psi research due to the widespread skepticism of the general scientific community. Without the money for research, it is almost impossible to create new and improved types of experiments that just might prove psi as lab-worthy.

Remote Viewing

Despite even more interest from the public and paranormal community, thanks in part to the entertainment industry's embracing of ESP and other psi phenomena in movies and television series, skeptics and scientists continued to claim little to no real evidence for psi existence. In spite of the skepticism from scientific circles, many governments, including the United States and the Soviet Union, have spent millions of dollars on experiments of their own. Most of that government sanctioned research involves clairvoyance, or the ability to see things happening at a distance—a widely reported phenomena more recently referred to as "remote viewing." Though much of the public's perception of clairvoyance comes from popular psychics such as Sylvia Browne, John Edwards, and others who claim to talk to the dead in order to help the living, or at the very least help you find your car keys, remote viewing has been at the heart of top-secret research funded by the CIA, NASA, the U.S. Navy, and U.S. Air Force since the early 1970s.

One such research project funded by the CIA in 1972 occurred at the Stanford Research Institute under the guidance of Russell Targ and Hal E. Puthoff. Both laser physicists, Targ and Puthoff met with personnel from the Office of Strategic Intelligence for the purpose of discussing paranormal experimentation. This led to an actual demonstration with a man who, in a laboratory setting, was able to alter the output signal of a magnetometer by simply focusing his attention on the interior of the instrument.

Targ and Puthoff continued their research for the CIA, using established psychics as "agents" to see if they could use their abilities to view locations or targets at a remote distance or time. Psychic and parapsychologist Ingo Swann was one of the best remote viewers the program found, along with Pat Price and Hella Hammid, who all took part in hundreds of experiments involving the remote viewing (RV) agent, drawing, or tracking the location of another subject at a randomly chosen location. Sometimes the subjects would be asked to just present the visual information of what they were seeing, and other times they were asked to draw the results. Price was especially successful at identifying sensitive military and government installations, as well as code names and layouts of these installations and eventually was asked to sign a secrecy agreement, thus becoming a covert employee of the CIA.

Swann also successfully identified sensitive underground installations, including an underground facility used by the National Security Agency in a remote West Virginia locale.

The program would continue to examine remote viewing as a military strategy for more than 20 years before it was made public in 1995. The use of remote viewers by the government for possible military purposes would create enough public interest, and outrage, to demand the release of these records, shining light on the seriousness of the subject matter. RV could be useful in identifying and locating important military targets, especially those of enemy nations, and could be especially useful in locating weapons of mass destruction.

The findings of these remote viewing experiments suggested that such RV abilities are latent in many people, some more than others, such as musical ability. The basic theory behind RV suggests that information is somehow transmitted from sender to receiver via the electromagnetic spectrum. Another theory suggests that remote viewers are accessing a sort of universal matrix of information that is available to anyone with the skill to tap into it.

But not every government study showed promise. In 1988, the National Research Council of the National Academy of Sciences released a report commissioned by the U.S. Army that examined controversial technologies for enhancing human performance, including parapsychology, where they deemed finding "no scientific justification from research conducted over a period of 130 years for the existence of parapsychological phenomena." Don't tell that to the millions of people who have experienced such phenomena!

Past Life Recall

One form of clairvoyance or remote viewing may involve insights into the past . . . past lives, that is. Past life recall is a widely reported phenomena that crosses religious and cultural boundaries, including boundaries of age. People from housewives to doctors, from small toddlers to senior citizens, have reported the recall of places, events, and people that they could not possibly have known in their present lives.

Past life recall, which has direct ties with religious or spiritual beliefs in reincarnation, involves signs such as recurrent dreams, extended déjà vu, knowledge of or ability in a field not acquired in present lifetime (foreign language, musical capability, and so on), strong and detailed memories of life in another time period, names of people not known in the present life, and even emotional recall of traumatic events not experienced in the present life.

Many past life recall experiences come out under hypnosis, but subjects, even young children, readily access past lives without prompting. Specific data such as names, dates, events, and locations can be corroborated, as in remote viewing, suggesting that subjects either truly did live a past life they can remember, or that they are possibly "channeling" the experience from an outside source, for instance a dead spirit. The problem scientists have with past life recall is the lack of a scientific method proving that what the subject is experiencing is indeed a past life, and not just memories of information gleaned subconsciously, or told to the subject by a relative or friend.

Near-Death and Out-of-Body Experiences

Déjà vu is often grouped in with past life recall but really has more to do with near-death and out-of-body experiences (NDEs and OBEs), which involve the mind's ability to travel beyond the boundaries of physical form (a more involved type of remote viewing perhaps). But while déjà vu, NDEs, and OBEs may have origins in natural, organic functions of specific parts of the brain, which would explain the common themes reported (as we will examine in detail in Part III of this book), they all may be part of something far more scientific than even scientists are willing to admit. NDEs and OBEs may in fact be triggered by the brain itself, but hundreds of studies done on patients in hospitals and clinics, including many young children, have led to the conclusion that NDEs and OBEs often involve the subject seeing things they could not possibly see from their body's location.

In a study done by Dr. Olaf Blanke, a neurologist at two Swiss University hospitals, the research team examined the brain activity of an epileptic woman who had been experiencing seizures for over a decade. They implanted electrodes and stimulated the angular gyrus area of her brain, close to the temporal lobe (the area Michael Persinger claimed was most effected by EM fields and that he himself stimulated with magnetic signals to elicit OBEs). The stimulation indeed caused the patient to have an experience of altered perception of her body, which is one of the "symptoms" of a NDE or OBE. This experiment, and many others, led the team to concede that they still did not fully understand the neurological mechanism responsible for OBEs. These experiments mirror studies done in the 1950s by Canadian neurosurgeon Wilder Penfield, who succeeded in eliciting OBEs using electrical stimulation, albeit on a different part of the brain known as the sylvian fissure, which divides the temporal lobes from the rest of the brain lengthwise.

But one thing the previously mentioned research teams could never seem to solve is the ability of OBE subjects to see things beyond their physical location. In other words, OBEs might be a more advanced form of remote viewing, which had already been proven to actually work in a lab setting. After all, weren't remote viewers mentally traveling to a distant location, even though in OBEs the journey seems much more specific and detailed? Obviously, the brain was creating the right environment for the experiences to occur, but could not be responsible for the end results. What could then?

Déjà vu, which is similar to a flashback of the present moment, certainly exists only in the mind of the person having the flashback. But what if déjà vu were more than just a strange electrochemical brain misfire? In fact, what if the recall of a situation that mirrors the one being currently experienced hinted at the possibility of cross-talk between parallel universes where we exist in many worlds, but cannot willingly access more than the one we have our "reality" in at any given time?

Even dreams, which are ordinarily thought of as our subconscious mind's way of communicating with us through symbols and images, may instead have a connection to parallel universes and hidden dimensions at the cutting edge of physics. Some parapsychologists even believe that Murphy's Law—everything that can go wrong will go wrong—may also have a scientific basis, suggesting that energy attracts like energy and that the human mind can indeed interact with inanimate objects and cause them to fail, usually at the most inopportune of times (like my computer!).

Precognition

Precognition, or the ability to see the future, is one of the most popular areas of psi, both with the public and the scientific community. If past life recall could help you understand why you hate olives or help you overcome your fear of mollusks, and if remote viewing of the present could help military strategists stop terrorism in its tracks and assist dim-bulbs in finding lost wallets, then imagine what seeing the future could do for anticipating trends and events in politics, business, entertainment, and even the outcome of the next Super Bowl!

Throughout history, there have been famous prophets and seers such as Malachai, St. John the Divine, Nostrodamus, and Edgar Cayce, whose predictions and visions of future times have intrigued skeptics and believers alike. Though open to interpretation, many of the predictions these and other seers have made have come true, and many have not. Many prophecies occur in dreams, such as the dream Abraham Lincoln had predicting his own demise only three days before his actual assassination at the hands of John Wilkes Booth. Lincoln reported the dream to author Ward Hill Lamon, who recounted it in his book *Recollections of Abraham Lincoln 1847–1865*. Lincoln's dream contained enough specifics that it is chilling to read it in light of what was to happen. Dreams, it seems, are more than just our anxiety-ridden subconscious working out its kinks.

I had a very frightening experience living in Los Angeles, which proved to me that precognitive ability is available to help us, if we listen to it. Unfortunately, I didn't. I was leaving at night from my North Hollywood apartment to go see Dr. Carl Sagan speak at an event sponsored by the Planetary Society in Pasadena. I was thrilled, having just read his novel, *Contact*. But as I headed for the door to leave, I was physically stopped by a force of energy I had never felt before, and my stomach turned. My hand gripped the doorknob, but the door would not open! I yelled to my husband, who just looked at me funny and told me to try the door again.

Despite feeling a pervasive sense of doom should I leave the apartment, I forced the door to open and left. I got in my car, still feeling as though something was telling me to turn back, but my desire to see Mr. Sagan was stronger than my intuition, and I headed for the freeway. It was an easy drive I'd made dozens of times, yet on this night, I somehow lost two hours of time that I have no recollection of, and ended up over 50 miles to the north, heading toward Sacramento on a totally different freeway.

What did I experience? To me, it was a premonition, a precognitive warning that I ignored, and paid the price for by experiencing a strange state that, to this day, I still have no solid memory of (and don't intend to find out under hypnosis, either!) and that got me lost far from home.

If seers are seeing the future, then where exactly are they seeing what they are seeing—and how? How are they accessing that information, if it hasn't even happened yet? And as many psychologists and dream theorists suggest, if dreams are nothing more than the gymnastics of the brain, how does that explain the brain's ability to occasionally cartwheel straight into the future?

Soul Energy

In *The Psychic in You*, author and intuitive psychic Jeffrey A. Wands asks the same question: "So where does this information come from? It comes from energy, the sort of energy that all human beings—living or dead—give off. You might call it soul energy . . ." He goes on to state that places and inanimate objects contain psychic energy, too. "Soul energy contains all kinds of information about the past, present and future. Our other senses can't perceive this energy, but it can be picked up by our 'sixth sense.'"

Edgar Cayce, known as the "sleeping prophet," knew all about this "soul energy" as well as its source. Cayce was born in 1877 in Kentucky. He grew up on a farm and first showed signs of an unusual talent when he would master his school lessons by sleeping with his books. After an illness that threatened to take away his voice, he began going into a hypnotic sleep state that allowed him to "see" a cure for his throat paralysis, and when he succeeded in curing himself, he began doing readings for others.

Cayce spent most of his adult life providing answers to questions for a variety of subjects, from the cause of someone's disease to the fate of the universe, and his transcripts provided the basis for several hundred books devoted to his work. The scope of his readings would eventually shift to more metaphysical subject matter, such as dreams, reincarnation, predictions of future events, and theories about the lost civilization of Atlantis.

But the most interesting aspect of his body of work was references to the Akashic Records or *The Book of Life*. Cayce believed that the Akashic Records contained the entire history of every living thing, every memory, event, and emotion, thought, deed, and intent that has ever occurred in history. This literal field of information and memory was similar to an infinite computer system that anyone could tap into at any time to access the past and present. And as some people believe, the future.

This field contains the "collective unconscious" in Jungian terms; the "Christ Consciousness" of the Christian mystics; the archetypes, symbols, mythologies, stories of cultures past and present; and even our dreams, hopes, and inspirations. Similar to the mind of God, this field, according to Cayce, could be accessed by anyone and used for guidance, direction, and knowledge. Similar to the Source of All Sources, the Akashic Records was considered a cosmic principle with a creative nature that gave birth, so to speak, to all physical matter, energy, spirit, and even thought. Philosopher Rudolf Steiner referred to it as a "spiritual world" that was as real as the physical, and claimed that the ability to perceive this other dimension of existence could allow man to "penetrate the eternal origins of the things which vanish with time . . ."

The Old and New Testaments of the Bible refer to a book upon which all has been written. The Book of Revelations specifically states that this book will be opened to judge the living and the dead. Ancient cultures such as the Sumerians, Babylonians, Assyrians, and Phoenicians all believed in the existence of a celestial tablet or book that contained the history of humanity, both physical and spiritual.

This idea that we all have a part of us that can access the past and the future, and even the other side of life itself, has a real basis in science thanks to recent discoveries in quantum physics and in the new science of brain/consciousness research. These discoveries suggest that the mind and the brain are actually two separate entities, with psi activity able to cross the bridge between them. What mystics and spiritual masters knew of centuries ago, men and women in lab coats are just now beginning to realize—there is more to the mind than meets the eye, or the brain. At a conference in Cambridge, England, in April of 2000, some 50-odd scientists from many disciplines were brought together to discuss the paranormal. Many of those present were physicists intrigued by the possible interactions between mind and matter.

One attendee, cosmologist Bernard Carr, told the Web site PhysicsWeb, "quantum mechanics, after all, is the first theory in physics in which the role of the observer has to be taken into account." Carr labeled psi phenomena in one of three ways: "pseudo-psychic phenomena" that may have a simple physical explanation; phenomena that may occur entirely within the mind; and real paranormal phenomena that suggests the mind interacting with the physical world. He defended physicists studying the paranormal by pointing out that physics itself is highly speculative in nature. "Some might say there is less evidence for superstrings than there is for ESP and at least we can try to replicate paranormal phenomena in the laboratory."

Studies by another attendee, Nobel Prize–winning physicist Brian D. Josephson, published in *Foundations of Physics* in 1991, would suggest even more connections between the quantum theory of non-locality and the ability of biological systems to create meaning from random patterns, something formal science was still unable to make sense of. These connections will be explored in Part III of this book, when we enter the realms of New Science and the powerful role the conscious observer plays in the perception of, and perhaps even construction of, reality.

Psychokinesis

Some of the more recent interest on behalf of both governments and the scientific community involves the potential manipulation of matter by the mind. Psychokinesis, or PK (once referred to as telekinesis), is basically "mind over matter." The word is derived from the Greek "psyche," meaning breath, life, or soul. PK manifests in a variety of ways, from levitating objects to bending spoons. Think Uri Geller and his bent flatware, the horror movie *Carrie,* or all those comic books you read as a kid where the superhero could stop oncoming locomotives with the wave of a hand. PK is often involved with poltergeist activity, although it can occur independent of it. J.B. Rhine, who initiated the era of ESP research, also looked into PK in a lab setting, finding the phenomenon similar to ESP, in that both were operating independently of space and time.

PK was once the activity of psychic mediums at séances, especially during the 1940s, but Rhine's research led to a new era of interest, and a separation of PK into two different categories: macro-PK, or observable events; and micro-PK, weak or slight effects not observable to the naked eye. American physicist Helmut Schmidt researched micro-PK in the latter part of the 1960s using a machine that operated on the random decay of radioactive particles. Subjects would be asked to try and mentally influence the flipping of coins in this "electronic" flipper, and this prototype led to the more complex random event generators (REGs), that have been producing significant results ever since, including some stunning tests done before and after the September 11th terrorist attacks in New York and Washington, DC, that hint at the power of collective thought to effect these intricate computers.

REGs look similar to a small black box, perhaps the size of a couple of cigarette packs placed side by side. Professor Robert Jahn of Princeton University was one of the first researchers to attempt experiments with REG. In the late 1970s, he literally hauled strangers off the street and asked them to concentrate on the number generator, asking them to try to mentally make it flip more heads than tails. The machine would generate two numbers, a one and a zero, or "heads and tails," in random sequences, and when the subjects were asked to make more heads appear than tails, the results were amazing, as again and again, regular folks like you and I could influence the machine with sheer mind power alone.

Dr. Roger Nelson, emeritus researcher at Princeton University, took up where Jahn's studies left off. He gathered 75 respected scientists from 40-odd nations to combine their time, energy, and talents to take part in the project (visit *http://noosphere.princeton.edu/* to learn more). The goal was to use REGs to detect whether humanity shared a single subconscious mind that could be tapped into, a mind that could not only effect matter, but predict the future. Nelson and his colleagues have added more Eggs, as the REGs have been named, for a total of 65, and the project was officially named the Global Consciousness Project.

Considered one of the longest-running and rigorous investigations of the paranormal, the GCP used the Internet to hook up 40-odd REGs across the globe to the lab computer of Nelson, who claims the graph on his computer looks similar to a flat line most of the time, except on those rare and unusual occasions, such as September 6, 1997, when billions mourned the loss of Princess Diana at her funeral service at Westminster Abbey. At that time, the graph shot upwards and recorded a massive shift in the number sequences of the globally positioned REGs. These deviations from the norm were big enough to let Nelson and his colleagues know they were on to something amazing, something that was duplicated each time there was an event of historic significance. The REGs could even detect mass public celebrations similar to those held each year on New Year's Eve.

Some of the most intriguing REG results occurred just before the terrorist attacks on U.S. soil on September 11, 2001, when an REG sensed the event hours before it occurred. The same results were evident right before the horrific Asian tsunami in 2004. The REGs significantly changed again during a global meditation held in conjunction with the Mukti-Maya unification period of the Mayan Calendar on June 1–2, 2005.

Scientists continue to rebuff attempts to study PK in a serious setting, but parapsychologists believe controlled experimentation will lead to wider acceptance of the phenomenon. One aspect of PK, called psychic teleportation, which is the movement of a person or thing from one place to another, seems serious enough to interest the U.S. Air Force, which reportedly commissioned a $7.5 million study titled "Teleportation Physics Study." Eric W. Davis of Warp Drive Metrics conducted the experiment.

The idea that the human mind can move an object from one location to another is still seen as science fiction by most mainstream scientists. But just as hundreds, if not thousands of reports of PK exist, including some that have been measured in a laboratory setting, teleportation is also not only possible, but in fact can and does conform to modern physics at the quantum level. Viennese physician Anton Zellinger, known as "Mr. Beam" by the media for his teleportation beliefs, has experimented with the teleportation of light particles. He claims to have first succeeded in 1997. In 2005, he teleported a light particle onto another light particle with no time delay over 600 meters under the Danube River. Other teams of physicists are successfully teleporting single atoms, but the teleportation of humans, at least in a laboratory setting, is still considered far beyond our understanding. At the quantum level, nothing seems impossible.

I've given you a basic overview of the world of the paranormal. I've hinted at the connections many of these phenomena may have with discoveries at the edge of the scientific envelope. Now, if you dare, come with me to a place where, in the immortal words of Lewis Carroll, author of *Alice's Adventures in Wonderland*, things just keep getting "curioser and curioser."

The Science

Something really strange happens when things get small. And the smaller things get, the stranger they become. Welcome to the world of the quantum.

No area of modern science is as bizarre, as mind-boggling, as seriously awe-inspiring as that of quantum physics, where all the laws that work so well on a larger scale simply fail to explain the behavior and dynamics of particles at the sub-atomic scale.

If you think ghosts, aliens, and ESP are "beyond normal," you might come to accept them as pretty bland in light of what we are about to learn. Because quantum physics defies logic, challenges common sense, and, in some cases, over-turns reality.

I am not a physicist, nor do I play one on TV, but I have been fascinated with quantum physics for decades, reading and absorbing any book I could get my hands on. Surprisingly, my interest in the paranormal and my continuing studies as a New Thought/Metaphysics minister would lead me even deeper into this realm of microscopic magnificence.

In the following chapters I will do my best to introduce you to the crazy, chaotic, and curious world of the teeniest, tiniest particles known to exist. I will try to avoid getting too technical, and I promise not to fog your brain with mathematical equations you will never understand. The great thing about quantum physics is that, as complicated as it really is, there is an elegant simplicity to the theories that allow ordinary people to grasp the concepts, even if they fail to get the math.

Because this book focuses on specific theories related to the paranormal and science of consciousness, I will present only those theories and some minor background, leaving you to your own devices to research more deeply on your own. Physics is way too big a subject to "quantumize" into just one section of one book.

By the time we move into Part III, you will hopefully have a basic understanding of, and a newfound appreciation for, a field of scientific study that is exploding with ideas, theories, and speculations beyond our wildest imaginings.

Classical Newtonian physics work relatively well for the big stuff, but it really does sweat the small stuff, as we are about to find out. Remember the old adage "less is more"? You bet your boson it is!

Quantum Physics 101:
Entangled Non-Local
Surfers Riding a Collapsing
Wave Function

*I ask you to look both ways. For the road to a knowledge of
the stars leads through the atom; and important knowledge
of the atom has been reached through the stars.*

—Arthur Eddington

Yah, dude, don't wipe out!

—A local surfer

Everybody knows the story of Sir Isaac Newton and the apple
that fell from the tree. But not everybody knows how this
simple incident involving a man and a piece of fiber-filled
fruit evolved into a classical branch of science that would one
day define the movement of stars and planets, and explain the
functions of force and matter.

Sir Isaac Newton was born in 1642 at Woolsthorpe Manor
in England. He attended Cambridge University in 1661, where
he was a noted student of mathematics. His achievements in
experimental investigation were innovative and creative, and led

97

to a new understanding of the science of mechanics. But he also delved into chemistry, the history of Western Civilization, and theology.

A well-rounded man of scientific genius, Newton would become the founding father of modern physical science, but forevermore would be associated with an apple. According to the story told in most science classes, Newton was spending time on his orchard and saw an apple fall. He conceived that the very same force that controlled the fall of the apple was what kept the moon in place. Immediately, he began making calculations on the amount of force needed to hold the moon in orbit as compared to the force pulling an object to the ground. These calculations evolved into others, involving the amount of centripetal force needed to hold a stone in a sling and the length of time it took to make a pendulum complete a full swing. Newton, you see, liked to calculate.

Not one to rest on his laurels, Newton corresponded with other scientists of his day to flesh out his theories, leading to his composition of a brief tract on mechanics titled *Principia Mathematica: Mathematical Principles of Natural Philosophy*. Divided into three books, Newton laid the foundation for the science of mechanics by identifying gravitation as the fundamental force controlling the motions of celestial bodies. He also posed theories involving fluids and the movement and motion of fluids, as well as the speed of sound waves. In the third book, he demonstrated the law of gravitation at work in the universe, pertaining to the revolution of planets and their satellites. Newton also explained tidal ebb and flow and computed the precession of the equinoxes from the exertion of solar and lunar forces.

This amazing contribution to science became known as classical Newtonian physics and was considered one of the greatest scientific achievements in history—until the latter part of the 19th century, when a revolution of sorts occurred in the world of physics that would take the emphasis off the cosmic scale and put it on the microscopic scale.

The Evolution of Physics

The foundations of quantum mechanics were laid in the latter part of the 19th century, with experiments that led to the discovery of black body energy in 1862 by Gustav Kirchhoff. Attempting to prove his theorem about black body radiation, Kirchhoff began a quest to determine the function missing from his equation, which proposed that the energy emitted by a black body depends on the temperature and frequency of the emitted energy.

It wasn't until 1900 that German physicist Max Planck identified the missing function and deduced the relationship between the energy and frequency of radiation. He based his revolutionary idea on the discovery that the energy emitted by a resonator could only take on discrete values, or "quanta." Planck also deduced that each and every quantum has a constant amount of action, and this became known as Planck's Constant, one of the two most important universal constants in physics (along with the speed of light). Planck's discovery marked a turning point in modern physics, and it also marked the true beginning of quantum physics, summarized in his book *Annalen der Physik*.

The word "quantum" originates from Latin, meaning "how much," and refers to discrete units assigned to certain physical quantities. This discovery that waves could be measured in small packets of "quanta" shined new light on the world of the atomic and subatomic systems, and soon science was buzzing with a new explosion of theories that classical physics quite simply could not keep up with. Where the physics of Newtonian science could provide accurate descriptions of the electromagnetic and gravitational forces on large objects, such as planets, quantum physics could explain the forces at work behind keeping an electron orbiting around the nucleus of an atom.

Albert Einstein would further Planck's work by showing that an electromagnetic wave, such as light, could be described by a new particle called a photon. This evolved into the theory of unity between subatomic particles and electromagnetic waves, called wave-particle duality, in which a particle and a wave are neither one nor the other, yet, mysteriously, can be both at the same time. Experiments would show that elementary particles such as an electron could exhibit particle-like behavior under some conditions, then exhibit wave-like behavior under other conditions.

Wave-particle duality is a central concept of quantum mechanics, and can be traced back to the 1600s when competing theories involving light were debated by Newton and Christiaan Huygens. Newton believed light was composed of tiny particles. Huygens theorized that light was composed of waves. One day they walked into each other and he got his chocolate in the other guy's peanut butter and . . . oh wait. Wrong story.

Thomas Young later solidified the belief that light was a wave with experiments in 1802, when the British physician and physicist conducted what are now referred to as the "double-slit" experiments. Young would force light from a single source to pass through the narrow slit of a board, and then force the same light to pass through two more narrow slits within a fraction of an inch of each other. The light from the two slits fell on a screen and Young discovered that the light beams spread apart and overlapped, and where they overlapped, bands of bright light alternated with bands of darkness. This led to his theory of light interference and established the wave nature of light. Young would use his new theory to propose that light waves were transverse (vibrating at right angles to direction of travel) rather than the originally assumed longitudinal (vibrating in the direction of travel). Young, an avid Egyptologist and no slacker, would also go on to help decipher the Rosetta Stone.

The later work of Einstein and others would prove that objects such as atoms have both particle and wave nature. This would include light, proving that every great theory is eventually shot down by another. Waveparticle duality would come to mean that quantum objects could behave both as waves and as particles, depending on when you observed them and how you observed them, and yet they could also display properties inconsistent with being both a wave and a particle. Confused yet? This duality defies common sense, but that is the nature of the quantum world.

In 1924, Einstein, who had already wowed the scientific community with his postulation of light energy associated with photons (for which he won the Nobel Prize in 1921), received a paper written by Satyendra Nath Bose. Acting as a referee to get the paper published, Einstein recognized the importance of Bose's proposal of different states for the photon. This paper led to the discovery of the Bose-Einstein condensate, a swarm of similar atoms cohering like a stationary laser beam existing at near zero temperatures and able to slow down the speed of light. Combined with the doctoral thesis of Louis de Broglie (extending the particle wave duality for light to all particles), the research of Erwin Schrödinger (introducing wave mechanics and developing an equation for the hydrogen atom), the success of Paul Dirac (unifying quantum physics with special relativity), and a host of other important theories and discoveries throughout the 1920s, this paper led to a literal reconstruction of understanding in the scientific world.

101 Particles, Particles Everywhere

Particles abound in physics. With the exception of the photon and graviton, all particles have distinct antiparticles. Most particles also have multiple spin states, and some have various color configurations (properties associated with their binding with gluons). The study of the basic elements of matter and the forces acting among them is called particle physics.

Leptons A family of particles consisting of muons.

Muons Heavier flavored leptons.

Taus Heaviest known leptons.

Neutrinos Uncharged, massless leptons.

Quarks Fundamental elementary particles with six types or flavors, including charm, strange, top, bottom, up, and down.

Electrons Negatively charged particles making up the outer shell of an atom.

Positrons Antielectrons. Positively charged antiparticles of electrons.

Hadron Particles made of quarks, including baryons and mesons.

Pions Lightest of mesons.

B-Mesons One of the heaviest mesons.

Protons Positively charged particles consisting of two up and one down quark that make up nucleus of atom.

Neutrons Neutral particles consisting of two down and one up quarks that make up nucleus of atom.

Bosons Particles associated with forces, including photons for electromagnetism, gluons for interactions of the strong force, and W and Z for weak interactions.

Gravitons Associated with gravitational force.

Fermions Matter particles (leptons and quarks are fermions).

Spooky Particles Refers to quantum entangled photons.

So much of quantum mechanics involves probability, that there is the likelihood or potential of finding a particle at a particular place at a particular time (say THAT 10 times quickly). The Heisenberg Uncertainty Principle was developed to handle this little detail, stating that there is an uncertainty in position of any subatomic particle, and that both the position and the momentum of a particle cannot be simultaneously known with total accuracy. Werner Heisenberg was a German physicist, born in 1901. He studied under other notable physicists such as Max Born, who in 1926 turned the classical physics world on its head when he abandoned causality in terms of quantum mechanics, and later Niels Bohr, who in 1922 won the Nobel Prize in Physics for his work on major laws of spectral lines.

Heisenberg published his theory of quantum mechanics in 1925 at the ripe, young age of 23. He would win the Nobel Prize in 1932 for his work, which led to the discovery of allotropic forms of hydrogen. But it was his principle of uncertainty that would take the scientific world by storm, stating that the determination of the position and momentum of any mobile particle will never be totally accurate, so we must therefore use the entire range of possible values to describe the particle's "probability distribution."

Thus, the world of the quantum would be founded on the back of probability. Generally, you cannot assign a definite value to an observable object in quantum mechanics, and must instead predict the probability of possible outcomes from measuring the observable.

Does your head hurt yet?

An eigenstate is a state associated with an observable object. Things that can be said to be in an eigenstate have a definite position, definite measurement value, and definite time of occurrence. A dog relieving itself on your shoe at 3 PM in your neighbor's driveway is a series of eigenstates.

But in the quantum world, things don't work that way. Instead, there is a range of possibilities and outcomes, and until the wave function is observed, and therefore collapsed, anything goes. This is called the Copenhagen Interpretation, which, to put it simply, states that nothing is really real until you observe it. Until the point of observation, reality is only a potentiality or a statistical formulation.

The Copenhagen Interpretation was formulated by Niels Bohr and Werner Heisenberg in 1927, and was an extension of the work of Nobel Prize–winning German physicist Max Born, who won the honor for his statistical interpretation of the wave function. Considered one of the most widely held views of mainstream scientists, the Copenhagen Interpretation holds that nature is inherently probabilistic and suggests a world where the intention or will of the observer has a direct influence upon the subjective reality.

Imagine a surfer riding atop a quantum wave, if you will. The wave function the surfer rides upon is only imaginary, existing in a sort of vibrational quasi-state until one particular thing happens—the surfer looks down at it. At the moment of observation, the wave function collapses and is manifested in our three-dimensional reality as, well, a gnarly wave, dude. But until the surfer bothered to look down at the wave, it remained in a spooky in-between state.

Thus, quantum mechanics exists upon the foundational belief that nothing can be determined about the true nature of a particle until it is observed. Spookier still, this implies a direct link between the particle (or object) being observed, and the observer, something we will get into more deeply in Part III of this book.

One of the classic experiments into wave function collapse involved not an apple, but a cat. Austrian physicist Erwin Schrödinger, a pioneer of quantum physics who, in the 1920s, focused on rules that govern the behavior of atoms and electrons by looking at how different elements radiated light, and the effects of radioactivity, decided to do a little experiment one day. His goal was to try and understand a complex theory called superposition, which stated that at the atomic level, because we cannot know what state an object is in until we observe it, the object remains in all possible states simultaneously . . . until we look at it. Then, as stated earlier, the wave function collapses and we "see" the object.

In other words, nothing is real until we look at it and make it so. Schrödinger took a live cat and put it in a box made of thick lead along with a Geiger counter with a tiny bit of radioactive substance. His assumption was this: there was the probability that in the course of one hour one of the atoms decays, but also, with equal probability, none, decays. If it did happen that an atom decayed, the counter tube would discharge and release a hammer, which would shatter a small flask of cyanide.

The cyanide would have poisoned the cat. But if there were no decay, the cat would be alive and rather annoyed at being stuck in a box.

But here's the kicker. Until the box was opened and the cat was observed, it would be in a superposition of states, both alive and dead, and everything in between.

And you thought UFOs were out of this world!

Now before you animal lovers get your Bose-Einstein Condensates in a tizzy, Schrödinger did not really use a live cat. In fact, this was a "thought experiment," which physicists working at the quantum level often resort to in order to try and sort out the possible rules governing the outrageously small (Einstein did it all the time).

Many-Worlds Interpretation

There is an opposite interpretation to the "nothing is real until you look at it" theory. The Many-Worlds Interpretation holds that everything is real, even when you don't look at it. According to the Interpretation, which was first proposed in 1957 by physicist Hugh Everett (although the term "many worlds" came from Bryce DeWitt, whose own formulation is often confused with Everett's), there are worlds that split off each time a potential choice exists. For example, in the cat-in-the-box experiment, if the atom decays, the cat dies. That is World Number One. But if the atom doesn't decay, the cat lives. That is World Number Two. In World Number Three the cat escapes, and so on and so on.

The Many-Worlds Interpretation opened the floodgates to the potential for an infinite number of worlds existing alongside our own. The wave function collapse leading to every possible outcome suggests a universe constantly splitting into another version of itself each and every time an opportunity presents an observer with a choice of all possible outcomes to a decision. Imagine the scope of such activity. Each time you decide to go get the mail, the universe splits to accommodate you both getting and NOT getting the mail. And should you run into a neighbor at the mailbox, the universe would have to split again to accommodate you both meeting and NOT speaking to said neighbor.

Talk about headaches! Just imagine a sea of ever-increasing universes popping up alongside the one you are standing in, and in each, you are making a choice other than the one you made in the one you are standing in.

That means there are infinite versions of you out there somewhere, doing an infinite number of things. As the observer of your own life experiment, each time you make an observation, you split into a number of copies, each observing one possible result and unaware of the other copies observing their results. The chain of copying continues in each universe that you and your copies exist in, resulting in a rapid and continuous splitting of real, but unobservable worlds.

Not to get all technical on you, but these worlds split because of the loss of coherency and the absence of interference between all the various elements of the superposition. Interactions with either the environment or a measuring device cause the wave functions of a system and the measuring device to become entangled. This is called "decoherence" and happens very rapidly on a macroscopic scale, which is probably why we cannot observe quantum affects in everyday objects.

Easier to comprehend, perhaps, is why we don't see these other worlds, or readily have access to them. The other worlds occupy the same space and time we do, but they do so in what many physicists describe as "Hilbert space," or hidden dimensions beyond our four recognized space/time dimensions.

The generally accepted Copenhagen Interpretation is still tops among many physicists, but more and more are entertaining the Many-Worlds, or "multiverse" theory, and naturally the concept remains a mainstay of science fiction novels and films. Because of Everett's formulation of the Many-Worlds Interpretation, other similar theories of the multiverse have entered the fray. There also exists a many-minds theory that postulates that the split occurs in the mind of the observer alone, meaning that the average mind is filled with an infinite number of alternate realities happening in other dimensions and universes parallel to our own.

Catching the Science Wave

I want to tell you a story about surfing in my hometown of San Diego, California. What does that have to do with quantum physics, you ask? Hold your hadrons, Grasshopper. I am about to explain.

Just about everybody in San Diego surfs. Well, not everybody. I don't surf. I have no sense of balance and I can't stand sand in my shorts. But a lot of people surf. On our most beautiful beaches we have what are called "locals." They know the water and the waves and consider the beach their own, and they frown upon outsiders or "nonlocals" using their stretch of beach.

These locals think they are the only surfers around, and that they are in no way connected with the tourists from North Carolina and Arizona who also want to try out a wave or two. Locals don't believe in sharing, or that what they do affects others at a distance. They live in their own little universe, riding the waves and baking under the hot sun of endless summers.

If Albert Einstein had been a surf rat, he would have been a local. Einstein, the genius behind the theories of relativity and special relativity, is still considered one of the most brilliant minds to ever grace the planet. Whole books have been written about his achievements, so for the sake of space, let's just say he helped develop the foundation of laws governing the universe, and that, to this day, most of those laws still stick.

In school, we all learned about Einstein's theories changing the face of science forever. His special theory of relativity proposed that distance and time are not absolute and depended instead on the motion of the observer. He introduced this theory in his 1905 paper "On the Electrodynamics of Moving Bodies," and expanded an earlier principle by including electromagnetism, which required the speed of light in a vacuum to be constant (186,282 miles per second) regardless of the motion of observers or the motion of the source of the light itself. Einstein, working off of earlier experiments by American scientists Albert Michelson and Edward Morley, confirmed that nothing could travel faster than the speed of light. Thus, the classical relativity of Galileo, which states that the laws of physics are the same in all uniformly moving frames of reference, got an updated facelift to include the laws governing light.

Einstein's Theory

In his theory of general relativity, Einstein proposed that gravity and motion could affect intervals of space and time. This geometrical theory postulates that the presence of mass and energy will curve space/time, affecting the path of free particles and light. The key to general relativity is the understanding of the "equivalence principle," which states that gravity pulling an object in one direction will be equivalent to the acceleration in the opposite direction. Interestingly enough, Einstein's predictions that light from a distant star would bend when passing the sun were confirmed in spectacular fashion in 1919, when a British expedition to West Africa observed a slight shift in position of stars near the sun during an eclipse. The light from the stars, as Einstein predicted, was bent as it passed by the sun. Here was direct evidence that space and time are warped. Gravity, it seems, doesn't attract or pull at all. Space pushes!

Scientists and historians called general relativity the first major new theory of relativity since Isaac Newton and the apple two centuries prior, and further predictions confirming general relativity would label Einstein as the greatest scientific mind of the last century, solidified even further by his association with the most famous scientific equation of all time:

LIGHTS ALL ASKEW IN THE HEAVENS

Men of Science More or Less Agog Over Results of Eclipse Observations.

EINSTEIN THEORY TRIUMPHS

Stars Not Where They Seemed or Were Calculated to be, but Nobody Need Worry.

A BOOK FOR 12 WISE MEN

No More in All the World Could Comprehend It, Said Einstein When His Daring Publishers Accepted It.

The world went "agog" over Einstein's theory, as the *New York Times* proclaimed.

$$E = mc2$$

This equation basically means, "energy equals mass times the speed of light squared" and tells us the energy that corresponds to a mass (m) at rest. When the mass disappears, this amount of energy must appear in another form. It also tells us that energy and mass are interchangeable, or that mass at rest is one particular form of energy. But then you remember all that from high school physics class, don't you?

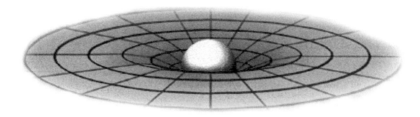

A body of mass curves or bends space.

Getting back to surfing, Einstein may have made some of the most fundamental discoveries of our understanding of the cosmos, and I certainly have only touched upon the tip of the iceberg of his contribution to physics and cosmology.

But when it came to surfing, Einstein wiped out. You see, Einstein didn't like some of the basic presumptions of quantum mechanics. They bothered him and had many strange characteristics that clashed with his concepts of how the universe worked on a cosmic scale.

In fact, Einstein thought some of the concepts of the quantum world were just plain "unheimlich," or "spooky" actions at a distance. The main thorn in Einstein's side was the issue of "nonlocality." According to the principle of nonlocality, two particles created at the same time, and becoming "entangled," can then be shot out into the universe in totally different directions. Then, if you change the state of one of the particles, it *instantaneously* changes the state of the corresponding entangled particle.

Einstein once told Niels Bohr, "God does not play dice with the universe." Bohr replied, "Stop telling God what to do!" To put it mildly, Einstein had a problem with the randomness of Heisenberg's Uncertainty Principle, which stated that you cannot accurately measure both velocity and position of a particle. Once you focus on one, you lose accuracy of the other. And now this spookiness of nonlocality, which meant that two particles could communicate *faster than the speed of light*, and that once you detected one particle, you could automatically detect the position of the other particle, really got Einstein frazzled. (Could that explain the hair?) So, in 1935, Einstein got together with scientists Boris Podolsky and Nathan

Rosen and created a list of objections to quantum mechanics, which became known as the "EPR paper" (the Einstein, Podolsky, Rosen Paradox).

Einstein and his collaborators found it simply unacceptable that two particles could communicate faster than light speed and keep track of each other over such vast distances. So they constructed a thought experiment that proved how ludicrous this quantum physics stuff really was. Their main argument was that quantum theory was incomplete and needed to be reformulated to observe light speed's limits, allow physical quantities to have well-defined values, and permit the accurate prediction of the outcome of a quantum event.

To paraphrase a famous courtroom utterance, if quantum theory couldn't make it fit, then you must a'quit.

There was just one problem. Einstein was wrong. After 30 years of debate, a theoretical physicist named John S. Bell, working at the CERN lab in Geneva, would conduct experiments on the measurements of states of polarization of light photons, and prove that, indeed, nonlocality was correct.

Bell's Theorem

Bell's Theorem basically asserted that *particles are connected on a level that is beyond time and space.* Theoretical physicist Henry Stapp stated that Bell's Theorem was "the most profound discovery of science." This theory that everything is connected to everything else sounds almost mystical . . . even metaphysical. But further experimentation proved nonlocality to be a reality that just would not go away.

So much for surfers who won't share their beach! Seems on the quantum level, every beach is connected to every other beach. And the same goes for the surfers, and the waves most of all.

The concept of entanglement was one of the few ideas that Einstein got wrong. If teeny, tiny subatomic particles remain entangled, and can affect each other at extremely vast distances, and do so infinitely fast, then what did that mean for the speed of light limitations that were so much a part of classical physics? We know from documented experiments that entangled particles must share a special kind of pairing of properties. In the case of photons, that could be polarization. For example, one photon could be polarized in the horizontal, the paired photon in the vertical; they just can't be polarized the same.

Theoretical physicist Michio Kaku refers to entanglement in his book, *Parallel Worlds: A Journey Through Creation, Higher Dimensions, and the Future of the Cosmos* in this creative way: "Entangled particles are somewhat like twins still joined by an umbilical cord (their wave function) which can be light years across. What happens to one member automatically affects the other . . . entangled pairs act as if they were a single object, although they may be separated by a large distance."

The photons are in a superposition, which as we learned earlier means they are in a "stateless" state until somebody comes along and measures them and collapses the wave function. An experimenter can then affect the atom housing the two photons in a way that forces it to shoot them out in opposite directions. Computers track the photons as one photon is forced through a vertical slit, and *instantaneously* the opposite photon freely takes on a horizontal polarization no matter where it is, and vice versa. If the first photon is forced through a horizontal slit, the opposite photon will "spin" vertical. Measuring the spin of one photon affects the measured spin of the other.

The quantum world operates on different laws than the larger scale of the cosmos, and the quest to unify the two worlds is the Holy Grail of physics. Where general relativity works for large bodies and celestial objects, it fails at describing the very small and the very fast. Quantum mechanics, though it provides accurate descriptions for phenomena that classical theories are unable to explain, fails to address the issues of objects larger than atoms and slower than the speed of light. General relativity, as well as classical physics, is deterministic, and quantum mechanics is undeterministic. And general relativity focuses on only one of the four fundamental forces in nature—gravity. Quantum mechanics relies on the other three—the electromagnetic, the strong nuclear force, and the weak nuclear force.

The fundamental forces have exceedingly different properties, and all play central roles in how the universe looks and operates. Gravitation is the most critical in respect to larger-scale cosmology, because it is long-ranged and able to act over longer distances, and because it supplies an attractor force between two pieces of matter.

Think "apple meets top of Newton's head on way to earth." The other three forces operate on either a shorter range or weaker scale. The unification of these strong, weak, and electromagnetic forces is what physicists refer to as the GUT—Grand Unified Theory. Or if you have a foot fetish, the TOE—Theory Of Everything. Theories adding gravity and attempting to unify the four forces into a single force are referred to as Superunified Theories, and remain, well, theoretical speculations for now.

There is light on the event horizon in the attempts of scientists to make sense of the universe. Stunning new theories may be the elusive Holy Grail of physics and cosmology long sought after by knights in white lab coats.

But first, we need to take a trip to some parallel worlds.

A Universe on Every Corner:
Or HELP! There's a Holographic Superstring on My Membrane!

Now, my own suspicion is that the universe is not only queerer than we suppose, but queerer than we can suppose . . .
                    —J.B.S. Haldane, *Possible Worlds and Other Essays*

Do I contradict myself? Very well then, I contradict myself. (I am large. I contain multitudes!)
                    —Walt Whitman, *Song of Myself*

Finding one theory that would unite general relativity with quantum mechanics has led to all kinds of amazing journeys for scientists willing to put on a suit of armor, mount a big steed, and go off in search of the cosmological Holy Grail. But instead of meeting up with dragons and trolls and beautiful damsels in distress, our knights in white lab coats meet up with parallel universes, 11 or more space/time dimensions, holograms, membranes, and strings so small they were worthless for sewing the gowns of fair maidens. King Arthur would have sent them to the tower without their supper!

The Big Bang

What may have started out as a great story device for science fiction novels and TV series ended up as a potentially comprehensive theory that just might be the One . . . or perhaps I should say, the TOE.

But before we get in between the TOE, we need to go back in time to the moment of universal birth known as the Big Bang. Actually, we need to go back to the moment just before the moment of the Big Bang, which many astronomers are now calling the "Big Splat."

Most, if not all, scientists studying the cosmos agree that our universe did indeed begin with a massive explosion—a fireball of tremendous power and heat that expanded on a timescale of one second and continues to expand, much as a balloon being filled with air. That our universe began as a random fluctuation in space, then went through a period of gonzospeed "inflation" before slowing down and cooling off long enough for matter and energy to be formed, is pretty much a given. In fact, as I write this book, scientists claim they can look as far back as the first trillionth of a second into the universe's existence, thanks to satellite images of the afterglow of the explosion.

But what happened right before the Big Bang? In other words, what banged into what to cause the Big Bang to go bang? Princeton University physicist Paul Steinhardt wondered where it all began, so he and Cambridge University physicist Neil Turok and Burt Ovrut of the University of Pennsylvania came up with a model of the universe that has no beginning and no end. Their universe operates on a cycle of expansion and contraction, and each expansion is its own Big Bang, giving birth to new universes at each noisy point of contact.

This theory is based upon the "ekpyrotic model" developed by Steinhardt and colleagues, and suggests the universe is made up of "branes," or threedimensional membrane-like sheet-worlds that are separated by a supposed fifth dimension that is no bigger than the hangnail of a flea. These membranes move like bed sheets, and every now and then, two of them meet in a slap or a "bang," thus creating another potential universe.

M-Theory

This is a simplified description of a much bigger theory many physicists think may be the great uniter. It's called the M-Theory, and the M refers to everything, including Mother, Master, Monster, Main, Miracle, and a host of other monikers that suggest this may be the closest thing to a TOE since the ankle bone made its debut. Edward Witten, the guy who actually proposed the existence of this new physical model, would, at a conference at USC in 1995, dub it "M-Theory," and he himself would suggest it could stand for "magic, mystery, or membrane, depending on your taste."

In order for M-Theory to work, scientists need to come to terms with the concept of an infinite number of parallel universes that exist alongside our own. The "other" brane alongside the one we exist on, and the one that banged into ours and gave us everything from life to reality television, is theoretically a "hidden" universe that is embedded in higher dimensional space.

Why is M-Theory so important, and why does it excite theoretical physicists? Because this theory alone brings together five similar, but conflicting theories involving superstrings, which just may be the most fundamental forms of matter in existence.

String Theory

String Theory has been around for decades, as a term to describe a group of related mathematical models of elementary particles and their interactions. Strings are actual one-dimensional objects that take the shape of open bits of line (or closed loops), that vibrate in varying modes. These amazingly tiny objects represent the only true fundamental particle.

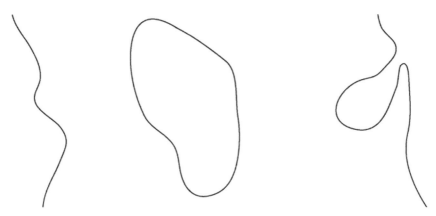

Strings come in open, closed, and looped forms. They can pinch off to form other strings.

(That is, of course, until someone finds something even smaller than strings!) String Theory actually began in 1921, with the development of the Kaluza-Klein theory, which stated that electromagnetism could be derived from gravity in a unified theory, but only in the presence of a fourth spatial dimension, or a fifth overall dimension (because we think of the fourth dimension as time).

Theodor Kaluza unified electromagnetic theory with Einstein's theory of general relativity and gravity from an idea based upon Hermann Minkowski, who had previously succeeded at adding a fourth spatial dimension to solve the space/time continuum. Kaluza, with Einstein's encouragement, published his theory, and in 1926, Oskar Klein went the next step, applying Kaluza's theory to quantum theory.

String Theory took off in the early 1970s, when particle theorists realized that theories developed in 1968 to describe the particle spectrum could also be used to describe the quantum mechanics behind oscillating strings. One of the key players was Gabriele Veneziano, a research fellow at CERN, the European particle accelerator lab. Veneziano observed that an obscure mathematical formula devised 200 years ago by Leonhard Euler called the "Euler Beta-Function" could describe many of the functions of the strong nuclear force. This led to more research by a number of physicists, who eventually realized that the nuclear interactions of particles modeled as one-dimensional "strings" rather than zero-dimensional "particles" could indeed fit the description of the Euler Beta-Function, just by adding the dimensional aspect of a string.

This was followed by the invention of supersymmetry, gravitons, supergravity, superstring theory, and a host of other developments that eventually led to a virtual revolution of String Theory in 1984, when strange anomalies of the theory were cleared up, allowing the theory to be accepted by the mainstream physics community.

I'm being very vague here, because you don't need to know all the details of String Theory to understand the importance of its implications or its eventual morphing into the mother of all theories: M-Theory. Suffice it to say that of all the theories theorized by theoretical theorists, the String Theory shows the most promise as a GUT, TOE, or precursor.

The Standard Model

Before the acceptance of String Theory, physicists struggled with the Standard Model of particle physics, which was a bunch of equations that identified the 12 basic particles that make up everything in the universe, and how they interact. Some physicists called the Standard Model the most successful theory of nature in history, but it has its problems. Those problems include the need to add more particles every now and then to explain some of the roles they play in higher-energy reactions, and the fact that it did not include gravity. Also, the Standard Model couldn't explain where all those particles came from.

According to Gordon Kane in his *Scientific American* article, "The Dawn of Physics Beyond the Standard Model," although the Standard Model needs to be extended, "its particles suffice to describe the everyday world (except for gravity) and almost all data collected by particle physics." Kane goes on to say that one of the questions that must be asked is, "If the Standard Model works so well, why must it be extended?" There are many mysteries the Standard Model cannot explain or account for, including the acceleration of the universe and whatever is causing it to accelerate; the possible "inflation" aspect of the universe at the moment of the Big Bang; the lack of equal parts of matter and antimatter in the universe; and the values of masses of leptons and quarks.

The Standard Model offers no answers, at least not yet. String Theory, on the other hand, would clear up some of the mysteries (but not all of them!).

Strings in Science

For now, this theory of strings and superstrings holds the most promise for bridging the gap that is the central conflict of modern theoretical physics—the incompatibility of relativity and quantum mechanics.

Strings represent the absolute rock-bottom level of fundamental particles. They are not made up of anything else, even though they do have a measured extent in space. They are it; you can't get any more fundamental. Well, that is until some theoretical physicist proves the existence of sub-strings, and trust me, there are rumors!

Strings have an infinite number of vibrating patterns, called "resonances," that not only correspond to different masses and elementary particle charges, but can also create them. The vibration pattern's energy is measured in amplitude and wavelength. A greater amplitude and wavelength means the particle has greater energy and mass. Moderate amplitude and wavelength corresponds to medium energy and mass, and lower amplitude and wavelength correspond to lower energy and mass.

Imagine the strings of a guitar, each one vibrating at a different frequency and making a different musical note. That's pretty much String Theory, minus all the technical stuff and math!

Newer research has led to the revelation of strings with more than one dimension, including those that operate in higher dimensions than our own four space/time ones. These are now referred to as "branes," with a one-brane being a one-dimensional string, and so on.

Originally, String Theory was just a possible explanation for the observable relationship between mass and spin of specific particles; the hadrons (includes proton and neutron) were left with a new theory all their own called Quantum Chromodynamics, and String Theory moved on to include particles known as gravitons. Gravitons have not been proven to exist yet, but theoretical physicists have long suspected they do, and so far only String Theory incorporates their unique particle qualities.

One of the biggest problems facing string theorists was the existence of five separate String Theories! This didn't include a sixth theory that only dealt with bosons (known brilliantly as "Bosonic String Theory"), but the other five theories all had to be reconciled. How could you truly have a TOE with five differing theories? Wouldn't that really be more like a BUM TOE? A Big Ugly Mess of a Theory of Everything?

The Five Theories

One thing the five theories do have in common is the necessity of 10 space/time dimensions. As physicists began realizing that these five separate theories were really just different ways of looking at the same theory, they did what they had to do and came up with *another* theory that would somehow incorporate the good, the bad, and the ugly, and make it all work together under one roof.

Michio Kaku, author of *Parallel Worlds*, states on his Web site, "String theorists are careful to point out that this does not prove the final correctness of the theory. Not by any means. That may take years or decades more. But it marks a most significant breakthrough that is already reshaping the entire field." Kaku adds, "In one stroke, M-Theory has solved many of the embarrassing features of the (string) theory, such as why we have 5 superstring theories."

Because each of the five string theories differ in symmetries, the addition of an 11th dimension in M-Theory helped to reconcile the varying symmetries, as well as the presence of what are called P-branes, or higher dimensional space/time objects, and D-branes, a special class of P-branes where the ends of the open string are localized on the brane (similar to a collective of strings in an excited state).

Believe it or not, as I write this, I am hearing of a 12-dimensional theory called F-Theory. F for Father. This bizarre theory has two time coordinates, and violates, according to Kaku, 12-dimensional relativity. Other physicists feel M-Theory, once finalized, may include no fixed dimensions at all.

I just don't understand. Can't these people settle on a theory and stick with it?

And what does all this have to do with parallel universes anyway? Patience, string people. I am getting to that.

M-Theory introduces the concept of membranes, as mentioned before. Our universe can be described as a three-dimensional membrane upon which we all exist. It could be just one of many smaller universes floating in a sea of other membranes in an even bigger universe. Physicist Lisa Randall of Harvard University, and author of *Warped Passages: Unraveling the Mysteries of the Universe's Hidden Dimensions*, writes about the possibility that our universe is a "three-brane" floating in higher dimensional space. And where theories of the past required many of those extra dimensions to be

When M-Worlds Collide

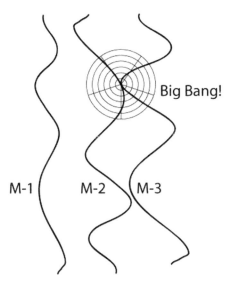

Big Bang!

M-1   M-2   M-3

(M = membrane)

The impact point of two membranes could have led to a Big Bang, and, our universe's birth.

tiny, curled up, and totally invisible, Randall is looking at the possibility that there could be dimensions of gigantic, even infinite size.

Randall's research was spurred by her realization that the other three forces are all pretty much the same strength, but gravity was weak in comparison. Yet gravity might just be a weak signal leaking out from a parallel universe or dimension less than a hair's breadth away from our own.

If our universe is just one "membrane" rippling in a sea of membranes, all wobbling their way through 11 dimensions, and if the Big Bang might have been the contact point of two rippling membranes, and if membranes bump into each other all the time, birthing more Big Bangs and more universes, then, to quote a movie poster from the 1970s, "We are not alone."

This would also mean that gravity is not as weak as we thought, but is actually strong at its point of origin, which could be a parallel membrane very close to our own. If some of this gravity leaks out into higher-dimensional space, it would make sense that we would only experience it in a weaker state.

Kaku believes membranes might also explain the existence of dark matter. If there were a parallel universe hovering very close to ours, it would be invisible to us, because it exists in another dimension we cannot access. But dark matter could very well be the end result of the presence of a parallel universe, which would affect the measurable properties of our own galaxy, thanks to the role played by gravity in attracting any large galaxy in the parallel universe.

M-Theory has its variations, the Ekpyrotic being one of them. Remember, that theory had two membranes colliding to create a new universe, and would certainly explain inflation—the flatness and uniformity of the universe. Other variations suggest universes can pinch off from one another, forming an ever-increasing foam of universes, the way a cell might split and divide. Another variation, specified by Robert Brandenberger and Cumrun Vafa, suggests the Big Bang is the result of the universe starting off as perfectly symmetrical, with its higher dimensions tightly wound up like a ball of rubber strings, only to break free from a collision between strings and anti-strings and expand rapidly outward.

No matter which theory is on the mark, the concept of parallel universes has taken off such as wildfire in the physics community, prompting books and lecture series on PBS, such as that of Brian Greene, author of

The Elegant Universe and Fabric of the Cosmos: Space, Time and the Texture of Reality. Greene has become somewhat of a household name for fans of PBS and the NOVA science series after hosting an informative and highly entertaining series based upon *The Elegant Universe.* Greene, similar to Kaku, focuses on String and M-Theories, and the necessity for 11 dimensions to accommodate both.

He cites the important work of Edward Witten toward advancing String Theory by pointing out the need for this 11th dimension, which unified the five separate string theories into one more cohesive, underlying theory: M-Theory.

Greene refers to the branes that strings exist upon as "braneworlds" and believes the higher-dimensional P-branes need not be as tiny as once suggested. In fact, similar to Lisa Randall, Greene feels that beyond the one-dimensional strings (visualize a power line), two-branes can exist on a large, two-dimensional surface (he suggests you think of a "ridiculously large drive-in movie screen"), and that three-branes can actually fill the three spatial dimensions. According to Green this means that we might right this very instant be "living in a three-brane," the latest "braneworld scenario" to be devised in the story of String/M-Theory.

Earlier variations of String Theory required extra dimensions to be tightly curled up into little balls, which would explain why we couldn't perceive those extra dimensions around us. But as Greene states, "it is not necessarily that the extra dimensions are extremely small. They could be big. We don't see them because of the way we see." He likens it to an ant walking on a lily pad, unaware of the water below the surface of a pond. We, too, could be floating on top of some vast higher-dimensional space.

We could also have plenty of company, too.

In an article in the May 2003 issue of *Scientific American*, titled "Parallel Universes," MIT professor and cosmologist Max Tegmark suggested there are infinite worlds with infinite copies of each one of us, identical in every respect except for one minor choice made that splits off into another universe, and thus, another copy. These copies, according to Tegmark, play out "every possible permutation" of our life choices. When I told my mom about parallel universes, she was thrilled by the fact that, in one of them, she was actually married to Johnny Depp . . . until I reminded her that this would also indicate that, in other universes, she was married to Hitler, Quasimodo, and a variety of circus freaks. (That didn't go over so well.)

Tegmark, and others, agree that the existence of other universes is implied by direct observations. Some of the theories they propose, called "multiverse hypotheses," include:

Open Multiverse An infinite number of, or collection of, universes that can exist in regions of space the same size as our observable universe. This theory suggests multiple Big Bangs occurring along a network of regions separated by huge distances, and relies on the concept of matter being distributed evenly across space.

Bubble Theory Suggests that our universe was just a bubble that arose from a multiverse, and that bubbles are rising up and creating more "baby" universes from the quantum foam of the parent universe. Kaku supports this "continual genesis" idea of multiple universes that bubble up from the cosmic foam.

Big Bounce An "oscillatory universe hypothesis" that suggests the universe goes through a series of infinite oscillations, starting with a Big Bang, and ending with a Big Bounce, where the universe collapses back inward from the gravitational pull of matter. This is linked to brane cosmology.

Many-Worlds Interpretation Hugh Everett's interpretation (discussed earlier) that suggests that universes that split off from one another all have the same physical laws and fundamental constants, but may exist in different states, which would account for the inability to "communicate" between universes. This is linked to the many-minds interpretation.

M-Theory 11-dimensional String Theory, suggesting universes are created by collisions between membranes. These universes will operate under different laws of physics, meaning anything goes!

String Landscape A type of String Theory suggesting that there are many ways to bring 10-dimensional String Theory to the 4-dimensional world we live in, with each way ending with a universe vastly different from our own.

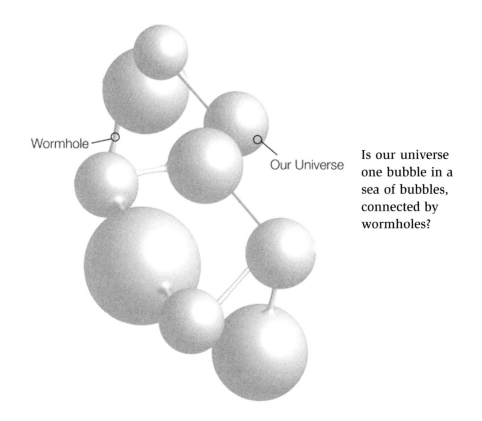

Wormhole

Our Universe

Is our universe one bubble in a sea of bubbles, connected by wormholes?

None of these theories has been proven, and all exist in the theoretical stage; critics point to a number of problems arising from such a complicated explanation for what we see and observe. Many point to the problems of including the observer in determining the existence of parallel universes, although others argue that observer effects must be a part of any scientific theory.

How Many Universes Are There?

Ockham's Razor, courtesy of 14th century sage William of Ockham, is also known as the "principle of parsimony," stating that one should never increase beyond what is necessary the number of entities required to explain something. In more modern terms: keep it simple, stupid.

But many cosmologists and theoretical physicists argue that the idea of parallel universes is actually an easier explanation for much observable data. This is easier than the idea of a one and only, is-that-all-there-is universe . . . the one we live in. This raises more questions, such as:

- Is our universe producing other universes?
- Can we be obliterated by another membrane crashing into us again?
- Can energy or matter move between universes in any state or form?
- Do universes compete for higher-dimensional resources?
- Do black holes play a role in producing new universes?
- Do all universes allow for intelligent life to emerge?

It is the last question that fascinates physicists. Referred to as the "anthropic principle," this theory suggests that, although there is an infinite sea of universes, perhaps not all have the potential to allow life. It would depend on whether or not each particular universe possessed the properties that were hospitable to life. This would include all the right elements for the formation of stars, galaxies, and planets, and the raw materials for life to spring forth.

The anthropic principle was formulated in the 1970s, by British astrophysicist Brandon Carter from Cambridge University. At a conference in Poland celebrating the 500th birthday of Nicolaus Copernicus, Carter presented a paper as a possible explanation for the observable fact that the fundamental constants of physics and chemistry must be perfectly fine-tuned for a universe to allow life to exist. These constants, which include the exact interactions necessary between the fundamental forces of gravity, electromagnetism, and the strong and weak nuclear forces, as well as the presence of water and the synthesis of carbon, had to be "just right" to produce life as we know it.

Some scientists have taken the anthropic principle to suggest the existence of God, or an intelligent mind plotting which universes have life and which don't. Others poo-poo the principle altogether, stating that it can't be tested, or that it holds no predictive power, despite its attractiveness to those who wish to believe that human beings were part of the equation right from the beginning.

Is our universe, then, perfectly suited to life by random chance? Some physicists and cosmologists don't think so. Sir Martin Reese expounds in

Our Cosmic Habitat that the multiverse could be something like an "off-the-rack clothes shop: If the shop has a large stock, we are not surprised to find one suit that fits. Likewise, if our universe is selected from a multiverse, its seemingly designed or fine-tuned features would not be surprising."

Author Tom Siegfried, in his book *Strange Matters: Undiscovered Ideas at the Frontiers of Space and Time*, suggests that we think of it like this: Each multiverse is a different physics laboratory with its own set of laws, and in some of those labs, under some of those laws, the raw materials for life can be cooked up. "Our universe has the physical constants it has, then, because it is the lab where life like us is possible."

Whether or not our universe had an actual beginning is of prime importance. As Siegfried points out, the theory of inflation, which states the universe was created from one tiny patch of space that "blew up" like a child's party balloon, inflating outward in roughly the same manner in all directions, must be considered when dealing with possible multiverses. If our universe could come from inflation, then so could others, in what could be an eternal inflation of bubble after bubble after bubble. This idea was actually outlined by the originator of the Inflation Theory itself, physicist Alan Guth, who came up with his concept in 1979 while at Cornell University.

Guth's theory of inflation stated that the universe expanded at a much faster rate than expected in the instant of the Big Bang. This period, called the inflationary epoch, was a consequence of the strong nuclear force breaking away from the weak nuclear force and the electromagnetic force, resulting in a phase transition that filled the universe with a kind of vacuum energy. The rate of expansion during this fraction of a second after the explosion was phenomenal, and suggests that everything in our observable universe was expanded from a volume of measurement only a few centimeters across. Guth's theory would itself be expanded with additional developments by Andrei Linde, Paul Steinhard, and Andy Albrecht.

Inflation Theory would account for the properties of some elementary particles that were not included in standard Big Bang models, and it could certainly account for many of the problems a Big Bang created for scientists; including the flatness of the universe due to the gigantic geometry of space, and the dilution of concentrations of "magnetic monopoles," which should have been present in huge numbers. It also cleared up the "horizon problem." The standard Big Bang model assumes the universe is both homogenous (which means matter is distributed evenly), and isotropic (which means it all looks the same in any direction). But in the standard model, different regions of space would have lacked the necessary time to produce the uniform distribution of matter and energy we see today. Inflation Theory states that the universe started out expanding slowly enough to allow for the needed "communication" between matter and energy interacting, thus creating the perfect conditions for the universe to "homogenize" itself before the real "inflation" began and the wild and rapid expansion period took place.

CNN News had a breaking story on March 16, 2006, about "the Big Bang's smoking gun." Physicists at the National Science Institute claim to have discovered absolute evidence of inflation by using the measurements of a space borne instrument called WMAP, the Wilkinson Microwave Anisotropy Probe, launched in 2001 by NASA. Researchers on the project examined variations of the microwave background over a part of the sky that was billions of light years across. They found new measurements involving patterns of light in the cosmic microwave background that prove our universe expanded to amazing proportions in the trillionth of a second after the Big Bang.

This exciting discovery will no doubt change how many cosmologists view the universe and its history. Inflation Theory may not be a theory at all, but a bona fide fact (unless someone comes along and discovers something new to disprove the proof) indeed includes the concept of "bubble universes" that could have inflated somewhere else, perhaps out of our range of perception because they are floating on membranes in another dimension.

Faster Than the Speed of Light?

In 1982, a research team at the University of Paris performed an experiment that would lead to the theory of entanglement. Led by physicist Alain Aspect, the team discovered that subatomic particles (under particular circumstances), could instantaneously communicate with each other no matter how far away they were, even at distances of billions of miles. They were able to show that these particles could actually do this at speeds faster than light, thus challenging Einstein's belief that nothing can move faster than the speed of light.

If particles, or any objects, could travel faster than light, they could also break the time barrier. This experiment led to more research as scientists scrambled to either disprove, or prove, Aspect's findings. One physicist, University of London's David Bohm, suggested that perhaps Aspect's experiment proved something else entirely: The universe we see is nothing but a hologram, albeit a huge and highly complex one.

A hologram is a three-dimensional picture that is made with a laser beam. Here is how it works: You pick an object, say an apple (in honor of Newton!) and bathe it in the light of Laser Beam #1. Then you take Laser Beam #2 and bounce it off of the reflected light of LB #1. This creates an interference pattern that can be captured on film. When the film is developed, you shine the light of Laser Beam #3 on it, and you see a three-dimensional image of an apple.

But that's not the end of the process, for if you were to cut that hologram of the apple in half, then shine one of your nifty lasers through each half, *each half would contain the image of a whole apple*! And spookier still, no matter how many times you divided those halves, you would still end up with an image of the whole apple. That's because any given part of a hologram contains all of the original information of the whole. You just can't slice an apple with a hologram!

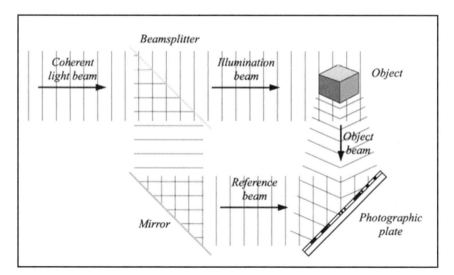

Holographic theory states that the universe is part of a greater, more fundamental whole image, as in a hologram.

Bohm used this type of imagery to suggest that particles could communicate with each other over great distances because they were not really separated after all. Instead, he applied the holographic principle to imply that particles at the deepest level of reality are not individuals, but parts of a greater, more fundamental whole.

The fact that these particles appear to be communicating faster than light could, in the holographic universe theory, be explained by this deeper level of reality, where we can only view parts of it from our perspective. This theory suggests that at this deeper level, everything is interconnected and that separateness does not exist, sounding an awful lot like mysticism and metaphysics. If the universe is indeed a projection of a larger image that exists somewhere else, that would go on to suggest that time, space, and even higher dimensions don't really exist, either, if everything is connected to everything else, at least not in the way we are used to viewing them. Our universe might be a three-dimensional world that is "painted" onto a two-dimensional surface of this greater reality.

Other Theories

In 1993, Dutch physicist Gerard 't Hooft of the University of Utrecht in the Netherlands proposed the Holographic Principle along with the co-inventor of String Theory, Leonard Susskind. This principle made two basic assertions that implied that all of the basic information (physics) in one part of the universe could be equivalent to some information (physics) defined on the boundary of that part of the universe. Think of it this way: All of the information in your body could be represented by your shadow cast upon a boundary or wall.

M-Theory suggests we are living on a four-dimensional brane afloat in a five-dimensional space/time. That higher dimension could be (or contain) the greater reality of the holographic image we perceive as our own brane existence, which is merely part of the greater reality. The holographic theory has gotten even more support with research into the physics of black holes, where the maximum entropy, or information content, of any region in space is defined by its surface area, rather than its volume. The term *holographic bound* refers to how much information can be contained in a particular region of space. Black hole thermodynamics are allowing physicists to make deductions of the absolute limits of the holographic bound.

A black hole could be the result of the surpassing of entropy beyond the holographic bound, whereby the sheer overload of information in one region of space could so exceed its ability to be contained within the surface area.

In his stunning and thought-provoking book *The Holographic Universe*, author Michael Talbot says, "Put another way, there is evidence to suggest that our world and everything in it—from snowflakes to maple trees to falling stars and spinning electrons—are also only ghostly images, projections from a level of reality so beyond our own it is literally beyond both space and time."

Talbot examines not only the work of David Bohm, Einstein protégé and quantum physicist extraordinaire, but the work of Stanford University neurophysiologist Karl Pribram, whose own research into the holographic nature of the brain eerily mirrors the holographic model of the universe. In fact, both Bohm and Pribram (working independently), came to the same startling conclusion: that the holographic model could possibly explain all of the mysterious phenomena in nature that no other theory could. This included telepathy, psychokinesis, mind over matter, and other psi phenomena. Pribram's work would suggest that the brain functions according to the holographic principle because the brain stores data, including memory, throughout its entire volume. We'll explore this more in Part III.

Talbot's book chronicles Bohm's process of discovery of the holographic theory, moving the physicist into areas of thought more suited toward metaphysicians. Talbot puts it rather imaginatively when he states, "Just as every portion of a hologram contains the image of the whole, every portion of the universe enfolds the whole." This idea of enfoldment could mean that the entire universe could indeed be held in a grain of sand . . . or the cell of a human being. In addition, a holographic universe would be one in which time ceases to exist in linear fashion. The past, present, and future would exist simultaneously.

In Bohm's holographic universe, matter does not exist independently from space, but is a part of space itself. This also applies to us at the human level. Talbot puts it into perspective by suggesting that our universe might just be a "mere hiccup in the greater scheme of things."

In other words, there are no locals on this beach. We all surf the same waves.

The holographic theory holds promise in the quest for a TOE. In *The Fabric of the Cosmos*, Brian Greene comments "I expect that regardless of where the search for the foundations of space and time may take us, regardless of modifications to string/M-theory that may be waiting for us around the bend, holography will continue to be a guiding concept." Greene believes that the holographic principle is one of the most likely theories to play a "dominant role in future research," noting as a strong piece of evidence the natural incorporation of the holographic principle into String Theory.

Only time, and plenty more theorizing, will tell if the universe we live in is but a tiny image of a greater masterpiece painted across the flat screen of reality.

Some scientists would liken the universe more to a computer. Information Theory proposes that everything in physical reality is pure information. Developed some 40-odd years ago as a means for maximizing the amount of information that can be transmitted over information channels, IT holds that the basis of reality is not the quantum, but the bit. Physicist John A. Wheeler is described by author Tom Siegfried in his book *The Bit and the Pendulum: From Quantum Physics to M-Theory—The New Physics of Information* as "perhaps the most distinguished evangelist for information theory." Wheeler, who also coined the term "black hole," refers to his brand of IT as "It from Bit," referring to the notion that, in his own words, "every physical quantity derives its ultimate significance from bits, binary yes-or-no indications . . ."

IT, in the cosmological sense, states that everything from energy to matter to space/time and even empty space itself has, at its foundation, an immaterial source of "bit soup," and that the posing of yes/no questions results in a response that suggests the universe is, in a sense, participatory. Wheeler stated in 1998 that the universe, similar to a computer, could indeed be based on this yes-or-no logic, and that "it is not unreasonable to imagine that information sits at the core of physics, just as it sits at the core of a computer."

The next question might then be, what is the universe computing, and are we part of the "output"? In *Programming the Universe: A Quantum Computer Scientist Takes On the Cosmos*, author and MIT professor Seth Lloyd suggests the universe is computing itself, and that as soon as the universe began, it began computing. The first patterns produced, according to Lloyd, were elementary and simple—particles and basic laws of physics.

But in time, the universe-as-computer processed more information in the form of intricate and complex patterns such as galaxies and stars. "The computational capability of the universe explains one of the great mysteries of nature," Lloyd states, "how complex systems such as living creatures can arise from fundamentally simple physical laws."

Lloyd argues that divine intervention, or the need for a "God" is not necessary in IT, which is one element of information theory that appeals to the purely scientific crowd, eager to remove any semblance of "intelligent design" from their final Theory of Everything. Similar to the movie *The Matrix*, IT puts humanity in the middle of a system that is controlled by the supercomputer, programmed from "seeds of complexity" that are based upon the laws of quantum and particle physics.

If indeed the universe is one big computer processing all of reality, it offers an explanation for the evolution of systems. Each time the supercomputer processes, it has access to more information from which to now build upon, so that the outcome is always evolving, always expanding. Ed Fredkin, considered an early pioneer of "digital physics," or as he now calls it, "digital philosophy," theorizes that atoms, electrons, and quarks are made up on binary units of information, and that the universe creates reality by endless repetition, taking in more and more information and transforming it, in a snowball effect, into greater measures of complexity.

Fredkin suggests that the universe is governed by a single algorithmic program, based upon the "recursive algorithms whose output is fed back into a universal computer as input." Thus, the next round of computation is even more powerful, having more information to work with. But it all comes down to the fundamental bits that make the it, and many physicists point to black hole entropy, which states that the total information contained in a black hole is proportional to the surface area of the event horizon, the point at which light can no longer pass, as a clue to the potential of IT as a final TOE. Up until now, physicists have been focused on field theory, but perhaps fields, from electromagnetic to gravitational to space/time, are not where the answer lies. Perhaps it lies in information exchanges that create the physical reality we see around us.

In his book *Decoding the Universe: How the New Science of Information is Explaining Everything in the Cosmos from Our Brains to Black Holes*, Charles Seife explains that quantum information theory, the study of qubits (quantum bits), is a hot area of research, and could provide some answers to a key source of conflict between general relativity and quantum physics. He focuses on entanglement. Einstein, he states, put a "speed limit" on the transmission of information between two objects, "yet quantum theory says that entangled particles *instantly* feel when their partners are measured." That these two particles are somehow communicating, and that this communication seems to move faster than the speed of light, boggles the minds of physicists such as IBM's Charles Bennett, who chose the term "teleportation" to describe the spooky action at a distance. Two teams of physicists in 1997 would conduct experiments involving the transfer of a qubit from one atom to another, proving that the spooky action at a distance is "quantum teleportation's mechanism for transmitting the quantum state of an atom onto another, the actual information on the atom can only travel from place to place at the speed of light." There was no way you could violate the ban on transmitting information faster than light-speed. At least not yet.

Information Theory also fits in nicely with the multiverse theory of Hugh Everett, especially where entanglement is observed. Charles Seife describes it by saying, "As information moves back and forth in the universe, it causes sheets to separate from each other, making the multiverse bubble and branch out." Information determines where the multiverse branches out, where it spreads away, where it sticks together. In other words, Seife summarizes, "information could be the force that shapes our cosmos."

In fact, IT fits in nicely with the idea of parallel universes, which can explain the many paradoxes of quantum mechanics, including spooky action at a distance. IT also suggests that black holes present an opportunity to understand how alternate universes form. What happens to black holes, which author Seife states, "are universes unto themselves," hints at the way information is preserved, and that it is proportional to the area of the event horizon, but it also seems to actually "live" on the surface of the event horizon, residing in two dimensions in a three-dimensional universe.

Kind of like a hologram. But the black hole appears to "record" the three-dimensional information (anything that gets beyond the event horizon), onto a two-dimensional surface area. Thus, the holographic universe theory is intimately linked with the Information Theory, and goes on to suggest that we humans are actually holograms ourselves—two-dimensional beings that exist in three-dimensional space.

Information is physical and must be stored somewhere, and black holes also lead to many ideas on how the universe might preserve and store information on a grander scale. Seife refers to finite spheres of information, including the one we live in, as "Hubble Bubbles," and states that, if the universe is infinite, our Hubble Bubble is one of many overlapping spheres (infinitely many, which means a large number of those could contain life as we know it, and life as we can't even imagine it). But because the sphere itself is finite, that would indicate that the information content is finite, which leads to a finite number of quantum states, and a finite number of wave functions . . .

This all leads to the possible extinction of our own little Hubble Bubble. Because the universe is expanding, it becomes harder and harder to preserve and duplicate the information required to sustain life. "As the universe reaches equilibrium, after a certain point the machines can chug away forever and they will not ever collect enough energy and shed enough entropy to give the civilization even another second of consciousness," Seife says.

But before you get depressed, consider that this won't happen for a long, long time, and by that time, we may have developed ways of moving between Hubble Bubbles. With human ingenuity, intelligence, and some amazingly fast computers, we just might do it within the next few centuries.

Aside from being a potential TOE, there are practical reasons for pursuing quantum information theory. In the early 1980s, physicists such as Paul Benioff, Richard Feynman, and David Deutsch began researching ways of blending quantum mechanics with IT, in order to one day create quantum computers that will make today's desktop models look like chalk and chalkboard by comparison. A quantum computer could solve problems much faster than the computers we know and love because it could achieve a superposition of states. Remember the Schrödinger experiment with the cat? The cat is in superposition until the box is opened and the cat is observed. But a quantum computer could possibly achieve a superposition of states for a specific problem, working on all possible outcomes at once, and come to a resolution much quicker.

Right now these cutting-edge computers are only possible in "thought experiments," but researchers are hard at work trying to turn the "it from bit" theory into a workable machine that would take a quantum leap in computing and processing information. Their goal is not just bigger, faster, better computers, though. For many of these scientists, quantum computers may hold the key to finally unlocking the mysterious code of quantum mechanics.

Information Theory offers some tantalizing ideas as to how the universe came into being, and what makes it work. It also may suggest how the universe will come to an end, but questions remain, questions that will hopefully be answered by a new wave of physicists eager to take the ideas of Wheeler and others and develop them further, such as William K. Wootters of Williams College. Wootters was a student of Wheeler's and is now pursuing the "it from bit" theory, even as many of his colleagues are pursuing String and M-Theory. Perhaps all these directions of study, including the holographic universe theory, will come together as individual pieces of a complex and profound puzzle, giving us a complete picture of reality.

No matter who finds the Holy Grail of cosmology, it will certainly take us to a whole new dimension of understanding the world around us.

Speaking of dimensions . . .

Up, Up, and Away!
Beyond the Fifth Dimension

When people thought the Earth was flat, they were wrong.
When people thought the Earth was spherical they were
wrong. But if you think that thinking the Earth is spherical
is just as wrong as thinking the Earth is flat, then your view
is wronger than both of them put together.
> —Isaac Asimov

All that we see or seem is but a dream within a dream.
> —Edgar Allen Poe

If we are surrounded by other worlds, many of which we might even exist in as copies of ourselves, then why can't we see them? And if we are but a holographic image of a higher dimensional reality, why can't we perceive that reality?

Logically, being able to perceive every universe we exist on would most likely drive us to insanity. It's hard enough to live one life; imagine having to juggle between an infinite number of lives! Clearly, our human brain could never handle the sheer information overload. Nor, as many scientists suggest, could it perceive the additional dimensions required for parallel universes to exist.

Dimensions

Ahhhh. Dimensions. You were wondering when I was going to get around to those. At the writing of this book, physicists seemed to agree that for parallel universes and M-Theory to even be possible, we would first have to accept the presence of 11, count 'em, 11 dimensions. Even the holographic principle requires a higher dimensional level of reality beyond the one we live, work, and play in.

The first thing we may want to do is make sure we understand the four-dimensional space/time we live in, because trying to move beyond that to envision additional dimensions is hard enough without first getting the 3-D Plus Time picture.

Imagine a straight line written across the page. That's a one-dimensional object. It has height perhaps, but no width or depth. Now imagine a child's long and pretty hair ribbon. That's a two-dimensional object, having both height and width. Now imagine a garden hose. That adds depth, so now we have our standard three-dimensional world.

A cube has three dimensions—height, width, and depth. You can also add on the fourth dimension of time that we live by, imagining the time it takes to get from one corner along the top of the cube to the opposite corner. Three-dimensional space can be broken down into three perpendicular axes: north-south, east-west, and up-down. A two-dimensional space would allow for only two axes, and a one-dimensional space, only one.

Imagine meeting a friend for breakfast at the IHOP on the fourth floor of the Eatemup Building on 4th and B Street at 10 AM. You would give your friend directions using three dimensions to locate the restaurant: the north-south and east-west location points and the floor it is on. The specified time you meet is the 4th dimension.

The battle goes on over whether the fourth dimension refers to time, which we will get to later, or a fourth spatial dimension. Some people call the fourth spatial dimension the "fifth dimension," including time as the fourth. This gets pretty confusing, especially for those of us who thought the fifth dimension was a 60s R&B group who sang about the dawning of the Age of Aquarius!

A fourth spatial dimension could look similar to a hypercube, or a "cube within a cube," although it is difficult for us three-dimensional creatures with three-dimensional brains to understand the hypercube. We can only visualize aspects of it.

In 1884, a man named Edwin Abbott published a short novel titled *Flatland: A Romance in Many Dimensions*. This quirky tale described the difficulties experienced by two-dimensional beings such as triangles, squares, circles, and polygons living on a plane when they got a visit from a being from a higher dimension—a three-dimensional sphere. In the story, the sphere reveals its true nature by raising its body through the flatland surface, giving the 2-D creatures the means by which they could perceive higher dimensions. But even then, they could not really perceive the higher dimensional object in a complete form, only certain aspects of it.

Similar methods of projection can be used by 4-D spatial objects to make themselves visible to our 3-D brains, and the rotating hypercube is one of them (do an online search for hypercubes to reveal animated examples). As the hypercube is rotated and shifted about a plane, we can see an aspect of its four-dimensionality, but our brains still may not fully grasp the whole bigger picture.

Maybe we don't see a fourth spatial dimension so easily because we don't need to. Our brains are programmed to react to objects up to three dimensions in space, so the evolutionary necessity of developing the visual acuity never arose. That doesn't mean it won't ever arise. In fact,

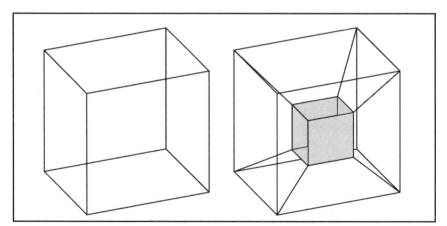

A three-dimensional cube and a four-dimensional hypercube.

we may have other ways of accessing, or sensing, higher dimensions. More on that in Part III, but imagine, if you will, this time-tested scenario. You decide to buy a bright lime green VW bug, thinking that, by golly, you are going to be the only human on Earth with a bright lime green VW bug. You even have to have one imported from Germany because you read about the rarity of bright lime green VW bugs on a Web site somewhere.

So, you get your nifty new car, hit the road, and—*there are dozens of bright lime green VW bugs everywhere you go!* Were they there all the time? Yes. Were you aware of them? No, not until you became personally invested in the awareness of them, namely because you threw down 20 grand for one.

This suggests that if we ever needed to see higher dimensional objects, because of a personal connection or awareness, the same principle might apply.

But until then, or until some computer whiz manages to create a program that easily projects objects beyond our 3-D world, we must settle for the world of the imagination.

One such imagining that has resulted in a computer model is the CalabiYau manifold, or as I like to call it, the Calabi-YOW! because it attempts to envision a six-dimensional spatial model. You can see some great 3-D projections of the Calabi-Yau on the net, but suffice it to say it reminds me of those Koosh Balls talk-show hostess Rosie O'Donnell used to throw at her audiences. Named after Eugenio Calabi and Shing-Tung Yau, the 6-D Calabi-Yau manifolds are, basically, shapes that satisfy the requirement of space for the six "unseen" spatial dimensions of string theory. These dimensions have (in the past) been imagined as curled up and so small as to be undetectable in size, which is described by the Calabi-Yau. Recently, theoretical physicists have begun wondering if these higher dimensions might instead be gigantic, even infinite in size. Or even spookier, that we may simply be living in a 3-D Plus Time subspace of a fuller universe, which is part of the braneworld theory.

Einstein came up with the first four dimensions we all know and love, and Theodor Kaluza suggested the presence of a fifth dimension in 1919, which colleague Oskar Klein later determined must be so compact as to be unobservable. Klein also envisioned this fifth dimension enfolded into normal space, sort of like a bubble inside of a more solid object. Now, String and M-Theories suggest the presence of 11 dimensions, with no end in sight as to how many more may be needed to properly describe the universe as we know it, and unify the four fundamental forces into one comprehensive TOE. (One version of string theory offers 26 dimensions, but let's not get too carried away. We can barely envision 11!)

When I think about the fact that other universes floating in higher dimensions exist at the tip of my very nose, it makes me wonder what would happen if I punched myself in the nose. Would I be committing "universicide?" It's a silly question, but one that many readers are no doubt asking right now. (Or not.)

Michio Kaku suggests in an article on his Web site titled "Meeting a Higher Dimensional Being," that if we were flung into hyperspace, we might see "a collection of spheres, blobs, and polyhedra which suddenly appear, constantly change shape and color, and then mysteriously disappear." He suggests higher dimensional people would have powers of a god in our eyes; moving through walls, seeing through solid objects, even vanishing and reappearing in a flash. Sound a lot like some of the paranormal phenomena described in Part I?

You're catching on!

Lisa Randall, in her book *Warped Passages* talks about gravity's potential to "leak" between dimensions, resulting in our own perception of gravity as a weak or diluted force, at least in comparison to the other three forces. If gravity can indeed leak off one membrane, through the higher dimension holding the branes in space, and to another nearby brane, thus resulting in what we experience when we sit under a tree full of heavy, ripe apples, could other forms of matter or energy also do likewise? Randall and colleague Raman Sundrum suggest other universes do exist, but in an extra spatial dimension that can only interact with ours gravitationally, if it interacts at all.

Most physicists will tell you, heck no. There is not even solid proof that gravity is breaching the interdimensional void. But in Part III, we'll take a look at some possibilities that might just make you wonder, and many of them are becoming more acceptable, if not downright plausible, in the scientific community. Right now, we need to take a closer look at the most mysterious dimension of all. Time.

Einstein's special theory of relativity shows that time behaves in many ways similar to the three dimensions of space. Length contracts as you increase speed, time expands at higher speeds, and you can measure time in a graph. So, we've come to refer to time as the fourth dimension.

Time As the Fourth Dimension

Our senses indicate that time flows. We cannot go back and fix the past. We cannot spring forward and shape the future. We are stuck in the present. There is just one problem.

There is no known law of physics that corresponds to the passage of time. That time "passes" may, in fact, be an illusion; an illusion perpetrated upon us by our conscious brain in an attempt to make sense of events in our lives and provide us with a way to measure our presence on earth.

Einstein's special theory points out that time is relative, and that two events occurring at the same time will appear to occur at different times depending upon the frame of reference of those viewing the events. The speed of light has to do with different references of time. If you are traveling near the speed of light, time slows. Traveling at the speed of light, time stands still. It's all relative.

So, if time is not a linear phenomenon, then what is it?

Some physicists refer to time as more of a "landscape" upon which the past, present, and future are all fixed. Theoretical physicist Paul Davies, in his February 2006 *Scientific American* article "That Mysterious Flow" suggests that, "later states of the world differ from earlier states that we still remember," giving the illusion of passing time. We, as conscious beings, register the "flow of time" as an arrow moving through space, so to speak. To the physicist, time is just an accurate measurement made by clocks.

Davies also suggests that past, present, and future must be "equally real," in a four-dimensional block universe with three spatial dimensions. The "world-line" or track through the 3-D space of events we perceive to be happening in the past, in the present, and in the future, have, in a grander sense, happened already and we are just processing our perception of them in a linear fashion. After all, an event is just a particular thing that happens at a particular time in a particular place. That is what makes sense to our brains, but it isn't the whole truth.

In classical physics, we can describe a series of past events by signifying everything that led up to them. But in the quantum world, there is no past, only probability, remember? The instant we measure a particle by observing it is preceded only by the probabilities of its location before the wave function collapsed. This suggests that the particle, up until the moment of observation, existed in all possible "pasts." This is referred to as the "sum over histories," which shows that a probability wave embodies within it all the possible pasts that could have preceded the moment of observation.

Talk about a new spin on history (pun intended)! In a very spooky experiment by physicist John Wheeler in the 1980s, photons were actually shown to adjust their behavior *in the past* based upon a future choice involving whether or not a photon detector had been turned on or off after the photon passed through a beam splitter. Called the "delayed choice" experiment, this suggested the photons, therefore, behaved as though they could predict whether or not the detector would be turned on or off *in the future*, and then they acted accordingly by adjusting their behavior *in the past*.

Another experiment, called the "quantum eraser experiment" showed that it is possible on the quantum level to actually erase the past, in principle, and another called the "delayed-choice quantum eraser" suggests the past can even be shaped. Remember the song by the Outsiders, "Time Won't Let Me?" Well, apparently if you're a photon, it will.

These complex experiments all involved subatomic particles, but the results are staggering nonetheless. Clearly, particles may not just be entangled in space, but in time. In Part III, we'll find out how traveling through time may not be "science fiction" after all, even for objects larger than a photon.

There may, in fact, be evidence of a second time dimension, one that is linked to M-Theory, thanks to one Michael Duff of Texas A&M, who ascertained that having two dimensions of time would help clear up some of the mathematical inconsistencies involving the need for 11 dimensions. F-Theory, the "father of all theories," which we discussed earlier, also suggests two dimension of time in addition to 10 of space. Cumrun Vafa, the father of the father of all theories, suggests that having this second time dimension makes sense in string theory as a means of making the "math work out," as Tom Siegfried writes in *Strange Matters*, but even Vafa admits it's all just theoretical at this point.

Perhaps the second of the two time dimensions might be suppressed, just as a holographic image of sorts. For those of us only able to perceive three spatial and one time dimensions, a second time dimension strains the brain, sounding more like a sci-fi novel plot device than something that might actually be feasible in a way that doesn't interfere with our known physical laws. As Astronomer Royal of Britain Sir Martin Reese states in *Our Cosmic Habitat*, if there were another dimension of time, we would need to develop a language with more tenses to describe what happens in it!

Tom Siegfried suggests, "In other words, the (3,1–3 spatial/1 time) signature of ordinary space/time is just the one that seems most convenient from the human viewpoint." He adds that nature may "encompass other signatures, some with more than one time dimension," all part of the grander concept, likening it to the five versions of String Theory that all turned out to be "the equivalent in the end."

String Theory, M-Theory, inflation, parallel universes, and hidden dimensions . . . all suggest exciting new potential for explaining how the world works. But before we go off in search of the missing links between the world of the known, and the world of the unknown, we need to make one last detour . . . toward zero.

The Field of All Possibility:
Zero Point Energy and the War on Gravity

Every great advance in science has issued from a new audacity of the imagination.
　　　　　　　　—John Dewey, *The Quest for Certainty*

We can lick gravity, but sometimes the paperwork is overwhelming.
　　　　　　　　—Werner Von Braun

Nothing exists in a vacuum. The concept of empty space has been shattered by the discovery of an infinite field of teeming activity, where tiny electromagnetic fields continuously fluctuate, even when temperatures reach absolute zero. A field where nothing, literally, is impossible because even at zero baseline values, there is something; quantum fluctuations or "jiggles" cannot be measured, yet permeate every inch of space.

A field that, if tapped into, might possibly produce enough energy to power the entire planet for a long, long time . . . and send a spaceship or two across the galaxy at near-light-speed to boot. The Zero Point Field.

CHAPTER 8

Zero Point Field

The ZPF is made up of Zero Point Energy (ZPE), a literal sea of energy that we swim in, as fish in the ocean, unaware of the vastness of our surroundings. ZPE was first suggested in the early years of quantum mechanics, when Paul Dirac theorized that the vacuum of space was instead filled with particles in negative energy states. These particles were predicted to materialize for brief periods, and exert a measurable force. This force was predicted in 1948 by Dutch physicist Hendrik B.G. Casimir. The Casimir Effect is a weak but measurable force between two separate objects, similar to two metallic plates hanging parallel to one another, which occurs due to the resonance in the space between the objects. This force can only be detected when the two plates are very close to one another, and the effect diminishes as the distance between the two plates increases. This force indicates a change in the electromagnetic field between the two plates. Lab experimentation has confirmed the Casimir Effect can also be a repulsive force.

The Casimir Effect proved the existence of ZPE, certainly in a scientific sense. As far back as 1911, Max Planck, Albert Einstein, and Otto Stern were researching ZPE, and in 1916, Walther Nernst formally proposed that empty space was filled with this field of zero-point electromagnetic radiation. Nobel Prize winner Willis Lamb was the first to measure the discrepancy between calculated and measured energy levels of hydrogen gas in an excited state, which lead to a greater understanding of vacuum field fluctuations and the development of quantum electrodynamics and the concept of Zero Point Energy.

In the Zero Point Field, particles pop in and out of existence, creating a "foam" of virtual particles that makes up empty space. Based upon the Heisenberg Uncertainty Principle, which states that the more accurately we can know the position of a moving particle, the less accurately we can measure its momentum. It also states that no quantum object can ever truly be completely at rest, the electromagnetic fluctuations of ZPE fill every corner of space, every nook and cranny, and are never at a state of absolute zero momentum, but instead vibrate at the most minute rate of oscillation allowable by the laws of quantum physics.

Thus, the tiny, residual "jiggle."

As it is the lowest state possible for energy to possess, the ZPF can only be visually detected in experiments such as the Casimir Effect. But were we to somehow magically remove all matter and energy above the zero-point state that exists in space, what would be left is the ZPF. Invisible to our eyes, but present everywhere. In her book *The Field: The Quest for the Secret Force of the Universe*, award-winning investigative journalist Lynne McTaggart chronicles the discovery of the ZPF and those involved with bringing this new science of the most fundamental source to the forefront of physics. McTaggart focuses on the work of laser researcher Hal Puthoff, who pioneered the remote viewing experiments discussed in Chapter 4, and his colleagues at the Stanford Research Institute. Puthoff had been interested in the Zero Point Field concept for many years, and was excited by the prospect of actually doing real research into what was thought of (at the time) as speculative.

Puthoff dove head first into the world of the ZPF, reading scientific literature, doing calculations, revisiting existing theories, and eventually coming to the conclusion that, according to McTaggart, "we and our universe live and breathe in what amounts to a sea of motion—a sea of quantum light." The sea is really a field, a matrix-like medium upon which the forces, like gravity or the electromagnetic force, move in ripples or waves.

This field in particular has been likened to the ether theory earlier proposed by James Clerk Maxwell, and later disproved by Albert Michelson and Edward Morley, who conducted a light experiment proving matter did not exist in a field of ether. Even Einstein, for a while, believed space was a void, until a 1911 experiment by Max Planck proved otherwise.

But no matter who or what preceded the concept of the ZPF, it is one theory that has the most promise as a potential TOE—Theory of Everything. As McTaggart says, "The Zero Point Field is a repository of all fields and all ground energy states and all virtual particles—a field of fields." Physicist Richard Feynman, in an attempt to describe the magnitude and power of the ZPF, stated that just one single cubic meter of space contained enough energy to boil the world's oceans.

Scientifically, the existence of a "field of all fields" or a "source energy" such as the ZPE was exciting enough. But for those who chose to look beyond the confines of strictly scientific thinking, the implications were nothing short of metaphysical. If this field existed in every bit of empty space between matter, then that meant that everything that existed in the universe was connected to everything else. By virtue of this field, reality is one big spider web with an infinite number of fine strands crisscrossing, intersecting, and creating a wholeness that extends throughout time and space.

Puthoff would soon discover that the ZPF operates on a self-generating feedback loop, meaning that "the fluctuations of the ZPF waves drive the motion of subatomic particles and that all the motion of all the particles of the universe in turn generates the Zero Point Field." Puthoff called this a "self-regenerating grand ground state of the universe."

The ZPF could explain a lot of the spookier aspects of quantum physics, such as nonlocality and the wave-particle duality. If everything were indeed connected in this sea of energy, then it would make sense that two particles could use the ZPF as the underlying mechanism for communicating at vast distances. Imagine that irritating guy in those cell phone commercials—"Can you hear me now?" The ZPF just might be the ultimate communications network!

Zero Point Field and Gravity

But the $64,000 question remains, could the ZPF have something to do with gravity? This weakest of the fundamental forces continuously proved to be the thorn in the side of physicists seeking to unify relativity and quantum mechanics, and even Puthoff wondered if the ZPF might be the bridge that would finally link the two. In 1968, Soviet physicist Andrei Sakharov suggested that gravity might be a residual effect of the interaction between objects, rather than the cause itself. This intrigued Puthoff, enough to attempt the further development of Sakharov's theory. Mathematically, Puthoff was able to show that the motion of particles in the ZPF could be consistent with the effects of gravity, proving in the process why gravity was such a weak force. His research detailed how the ZPF could be the missing link in the quest to bring together all of the four fundamental forces into one TOE.

McTaggart documents how Puthoff put his theory before the scientific public, to lukewarm receptions, and then dove into more research at the home of a lab engineer named Ken Shoulders. Their work focused on "condensed charge technology," and they designed some prototype gadgets that could use the technology when factoring in the ZPF. Even the Pentagon would jump on the "condensed charge technology" bandwagon, confirming Puthoff and Shoulder's work, but it would be almost two decades later when the next revolution of the ZPF would occur.

Puthoff would begin working with physicist and applied mathematician Alfonso Rueda and astrophysicist Bernie Haisch on the potential connection between inertia and the ZPF, which resulted in a lengthy paper published in 1994 by the *Physical Review*. That paper would be called "a landmark" by science writer Arthur C. Clarke. It suggests that inertia and gravitation are both electromagnetic phenomena that result from an interaction with the ZPF.

This remarkable paper opened the door to a world of possibility, namely the world of "antigravity" and the possibility of space travel by linking inertia, mass, and gravity to the ZPF. As McTaggart stated in her book, "If you could extract energy from the Zero Point Field wherever you are in the universe, you wouldn't have to carry fuel with you, but could just set sail in space and tap into the Zero Point Field . . ." She called it a "universal wind" that could, as Puthoff and others hoped, one day send space vehicles sailing into the deepest regions of the universe, and do it cheaply.

It Keeps Going and Going . . .

The ZPF is estimated to be massive, even infinite, and to exceed nuclear energy densities, meaning that just a small amount of ZPE could provide a whole lot of fuel. In fact, many scientists look to the ZPF as a potential source of extracting free and unlimited energy that may one day power homes, cars, and businesses. Science fiction novels and television shows already hype the ZPF as a powerful source for creating everything from *Star Trek*'s quantum torpedoes to *Stargate SG-1*'s modules made in the field that allow for intergalactic space travel.

Although currently we can only measure minute amounts of ZPE levels, physicists such as Puthoff believe we can achieve the technology to one day tap the field in much bigger ways. The aerospace industry seems to believe we can achieve this as well. A March 2004 article in *Aviation Week and Space Technology* titled "To the Stars" stated that two large aerospace companies, and one U.S. Defense Department agency are betting on ZPE, launching bold research projects exploring the potential energy source. Puthoff stated in the article that the potential is "practically limitless, way beyond what can be conceived." But he points to the need to first design a way to extract the energy from the Zero Point Field, a process that—as of yet—remains utterly inefficient at producing more energy than "a butterfly's wing."

There is also a yet-to-be-found catalyst that would "ignite the ZPE process."

This "new physics" of the Zero Point Field could one day take us to the nearest planet in a matter of weeks instead of years. As a method of propulsion, the sky is literally the limit, thus the intense interest in the ZPF by NASA and both government and private industries.

Zero Point Field in the Real World

But intense interest in the ZPF and its potential for powering spacecraft was not limited to U.S. agencies, nor was it limited to the last three decades. In *The Hunt for Zero Point*, author and *Jane's Defence Weekly* aviation editor Nick Cook documents the Nazis' intense interest in anti-gravity and Zero Point Energy. This potentially limitless source of power intrigued scientists in Nazi Germany, who were later brought over to live in the United States as part of Operation Paperclip. These scientists believed in ZPE as not just an energy source for fueling rockets and planes, but as the potential material for a powerful bomb.

Cook chronicles the quest to control gravity and take to the stars at speeds near or surpassing that of light. Military, aerospace, and corporate interest in ZPE has been heated since the 1940s on both sides of the Atlantic, beginning with the concepts of "electrogravitic lift" of T. Townsend Brown, an inventor who, in 1929, wrote a paper called "How I Control Gravitation" to accompany his own creation—an electrical condenser he called the "Gravitor." This device was a type of motor that utilized the principles of electro-gravitation, and led to Brown developing the ideal shape for electro-gravitational lift—the disc. This was in the 1920s, when the aviation industry was still trying to get a fighter plane to fly faster than 160 miles per hour.

Brown's research would lay the foundation for his later work with the Naval Research Laboratory, where he would be assigned to work on experiments with acoustics and minesweeping. But he would, during that time, invent a method for canceling a ship's magnetic field, a critical element in wartime, and would later be linked to the notorious Philadelphia Experiment, which supposedly involved the disappearance of a naval warship and its crew into another dimension.

Brown went to demonstrate his Gravitor and flying discs to military officials eager to grasp the potential of defying gravity. Eventually he established his own research foundation, continuing to pitch to the military his amazing disc technology.

Even the Nazis and their brilliant scientists were at work on antigravity technology, which Cook believes might account for the "foo fighters" so often spotted by Allied pilots during World War II, and of course, the United States was forced to keep pace, doing their own black-budgeted research. Other nations would step into the fray, with Russian, Finnish, and British scientists all searching for a method of not just controlling gravity, but overcoming it altogether.

Much of the groundwork into antigravity and ZPE had been laid by T. Townsend Brown, and German scientists working for the Third Reich (both voluntarily and involuntarily), such as Viktor Schauberger, who had built an unconventional machine while interned in a concentration camp in the early 1940s that generated lift, dubbed the "fleigende scheibe," or "flying saucer." The craft would later be dubbed the "Repulsine," and would be one of many prototypes created and tested under the Nazi regime.

As we saw in the first chapter of this book, Avro Canada would decades later develop, test, and reveal to the public their own flying saucer, the Avrocar, but that would be a dismal failure as a potential for future space flight. It could barely get off the ground.

Zero Point Field and Science

ZPE research would fall into a kind of black hole of its own for the next few decades, with little public information and even less government admittance that it was a serious pursuit. But documents recently declassified and investigative reporting such as Cook's reveal a continuing interest at NASA and other government agencies, all of which were spearheading (including financially supporting) the work of various researchers. Russian intelligence agents also showed interest in the lifter technology of Viktor Schauberger, suggesting their own ongoing black program into antigravity.

UFO sightings would be linked with the disc technology of the Nazis, and when America made their power grab of the German technology after the war, many ufologists would wonder just how many UFOs were from "out there," and how many were from "down here," like the recently declassified Project Silverbug, a supersonic saucer developed by the

United States Air Force. Silverbug was rather conventional, being a jetpowered vehicle, but for the Americans, it was an attempt to develop prototypes closer to what the Nazis had been developing before the end of the war.

Lifter technology, levitation, and antigravity all got a real kick in the pants with the work of a man named John Hutchinson. Born in 1945 in North Vancouver, Hutchinson showed an intense interest in machines from an early age. He studied the theories of Einstein, but was obsessed with the theories of Nikola Tesla, the Serbian electrical engineer and brilliant inventor who would literally introduce the world to the method of alternating current (AC), and eventually, the Tesla Coil, which became the basis for radios, televisions, and wireless communication devices.

Tesla, although brilliant and bold, would never achieve the respect of say, Thomas Edison, but his experiments would push the envelope of the possible, first suggesting that Earth could become a powerful conductor of electricity. He would also be linked to wilder theories and experiments involving "death rays."

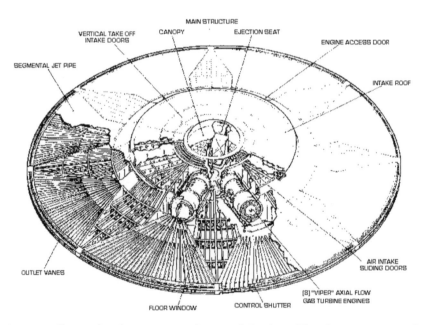

German disc technology centered around Project Silverbug, a manmade UFO with antigravity propulsion technology.

The Hutchinson Effect

Hutchinson made his own laboratory as a teenager, filling it with Tesla devices and all kinds of machines. But in 1979, he would be sitting in his lab when, out of nowhere, he was smacked on the shoulder with a piece of metal. With some tweaking of his equipment, and some further experimentation, he discovered what would be known as the "Hutchinson Effect," the ability of ordinary household items to levitate, bend, break, move, and explode due to a "lift and disruption" that was occurring in his basement lab.

The Hutchinson Effect, which had been captured on videotape and film, seemed to be caused by some invisible energy force present in the lab under specific circumstances. It didn't happen all the time, but when it did, the proverbial you-know-what hit the fan, along with anything else in the lab that wasn't nailed down.

Many theories abound as to the origin of the effect, from psychokinesis to the Zero Point Field to the presence of opposing electromagnetic fields. But one thing was certain. John Hutchinson was onto something, and even the Pentagon was interested in finding out what it was, and if it could be duplicated. Eventually they would move Hutchinson out of his basement to a nice warehouse with tight security and classified status.

Hutchinson would duplicate his effect for Pentagon bigwigs, and even claim to have created pockets of time dilation, altering space/time within target areas. There seemed to be one problem though. The effect would only occur when Hutchinson was present. Eventually, Boeing and McDonnell Douglas, two major American Airline companies would fund some experiments in the 1980s for the Hutchinson effect, but eventually Hutchinson would go off on his own. His effect would be featured on many television programs on the Discovery Channel, TLC, and on Japanese television, and he even sells videos of the effect if you are willing to fork over 100 bucks.

New Findings in Modern Technology

Into the 1990s and the new millennium, antigravity and lifter technology continued to capture the imaginations of researchers, physicists, and corporations eager to be the first to patent and market the first "freeenergy" generators, machines, and vehicles. *Wired* magazine sponsored a photo/video shoot of a lifter experiment courtesy of American Antigravity and is viewable on its Web site. American Antigravity's Lifter project, under the guidance of developer and UNIX programmer Tim Ventura, has gained plenty of media exposure and is one of many independent research programs pursuing lifter technology.

Lifters are cheap, lightweight crafts made of aluminum foil, balsa wood, and thin wire, which can defy gravity and levitate using little more than a ground-based power supply. These lifters have no wings or propellers, and are tethered to keep them from flying too high once the power source is switched on. Lifter experiments are all over the Internet, complete with photos and videos, as inventors compete with each other over how much mass their lifter can, well, "lift." Ventura and his team are working to convert electrical current into a force that can lift more than just a strange contraption made of foil and balsa wood, though. They hope to one day develop the technology to lift planes, rockets, and even move earthbound vehicles such as trains and ships.

The antigravity lifter used in the *Wired* magazine experiment. Courtesy of Tim Ventura/AmericanAntigravity.com

In early 2006, Congress even decided to earmark funds to further study lifter technology.

"Table-top antigravity" is hotter than ever as researchers attempt to create everything from levitating superconducting ceramic disks to asymmetrical capacitors to lab-tested devices that tap into "energetic aether."

The PSIence

Do we need more reality? We've already got so much.
—Ralph Abraham in *The Evolutionary Mind*

We have actually touched the Borderland where matter and force seem to merge into one another, the shadowy realm between the known and the unknown . . .
—Sir William Crookes, 1879

French author and UFO researcher Aime Michel once said that whenever something of a superhuman nature manifests itself, the apparently absurd is what you should expect. Many of my first encounters with the apparently absurd came as a child, when I began to have "experiences" of other entities and creatures that shared my space. These imaginary friends were real, at least to me, and occupied my waking and sleeping consciousness with a depth even my mom couldn't fathom when she insisted they were figments of my imagination.

I knew, despite the protestations of those far older and, supposedly wiser, that they were real (including one particularly snappy alligator who carried a briefcase).

Little could my mother, or I, have predicted that some three decades later, physicists on the cutting edge of their field would be talking about figments of imagination as realities in parallel universes; realities that, on occasion, slip into that crossroads area and choose the path not taken, the path that would lead them smack into a head-on collision with human consciousness.

In previous chapters, we've seen examples of UFO and paranormal phenomena that demand an explanation involving not this world, but others interconnected with our own. From Jacques Vallee's massive research into the interdimensional origin of UFOs and aliens; to Edgar Cayce's readings from the Akashic Records of universal memory; to the many cases of poltergeist activity, ghostly hauntings, electromagnetic anomalies, and bizarre creatures that slip in and out of existence, all these experiences hint at more than just a terrestrial, or even extra-terrestrial, origin. In fact, they seem to hint at something "para-terrestrial."

We know from earlier chapters that Bell's Theorem demonstrates proof of nonlocality, which theorizes that everything in the universe is intimately connected in ways that cannot be perceived from a three-dimensional perspective.

Hal Puthoff also believed that "we and all the matter of the universe are literally connected to the farthest reaches of the cosmos through the Zero Point Field waves of the grandest dimensions." Knowing that vacuum fluctuations occurred in the ZPF, Puthoff theorized that levitation could indeed occur when the ZPF was harnessed at will.

Hugh Everett's "many worlds" theory is of great significance to parapsychologists and physicists who are coming to believe that there are ways to move between these separate universes. Thus, a person who died in one world could still manifest in the next, or an alien ship that took off in Universe A could find its way to a McDonald's drive-through in Universe D.

But how could any of the scientific theories examined in Part II, including String Theory, M-Theory, the holographic model and many-worlds, even the Zero Point Field, *physically* account for ghosts, mythological monsters, ships that vanish at sea, missing planes, mysterious human disappearances, poltergeists, UFOs and aliens, missing time, Shamanic travel, near-death experiences, angelic visions, clairvoyance,

ESP, PK, and all sorts of spectral creatures that roam the countryside from England and New Jersey?

If other universes exist in other dimensions, where is the window between them and us, and how could paranormal entities and energies break on through from one side to another?

And could we one day do the same?

In other words, how *do* those UFOs get from there to here? And what could our own consciousness have to do with it all?

In the movie *The Matrix*, Morpheus offers Neo two pills—a red one and a blue one. He tells Neo that the blue pill will lead him back to the safety of the known and that he will continue to believe what he has always believed.

But the red pill . . . ahhh, the red pill. If Neo chooses the red pill, he will be taken deeper and deeper down the rabbit hole to a place where anything, and everything, is possible. Including the truth about the reality of the matrix itself.

You Can Get Here from There:

Wormholes, Time Warps, and Interdimensional Truck Stops

Space isn't remote at all. It's only an hour's drive away if your car could go straight upwards.

—Fred Hoyle, *Observer U.K.*, Sept. 9, 1979

In order to more fully understand this reality, we must take into account other dimensions of a broader reality.

—John A. Wheeler

Time is God's way of keeping everything from happening at once.

—Unknown

The fastest path between two points is usually a straight line, but in curved space/time, that simple line is not always possible. For a spacecraft to travel the jaw-dropping distances between stars and galaxies, the amount of fuel needed would be extraordinary, if not impossible. And if your destination lies in another universe outright, one parallel to our own, then it is imperative you do something most of us do when faced with a long, tedious trip. You need to find a shortcut.

Wormholes

Most scientists agree that black holes exist, simply because the evidence pointing to their presence in deep space is undeniable. Most scientists also agree that the possibility of parallel worlds is becoming an increasing probability. Some scientists think that there are ways of getting here from there, and they call them wormholes.

Imagine, if you will, Newton and the apple that led to the discovery of gravity. If a worm wanted to get from point A on the apple's surface to point B, he could slither all around the curvature of the apple; or he could do what worms do best and burrow straight through to the other side. (This is why you should never bite into an apple with holes in it.)

In 1935, Einstein collaborated with Israeli physicist Nathan Rosen on a paper that proved the singularity of a black hole could be a bridge connecting two universes. This bridge became known as the Einstein-Rosen Bridge, but most physicists call it a wormhole. Some people call it a rabbit hole because it was English mathematician Charles Dodgson, known better by his pen name, Lewis Carroll, who wrote a story about a curious girl named Alice who follows a white rabbit down a hole into a parallel world where the laws of physics are turned upside down. *Alice's*

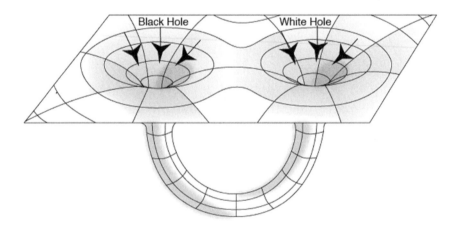

A typical wormhole, with a black hole as entry point, and a white hole as exit point.

Adventures in Wonderland was written in 1865, when the ideas of parallel worlds and alternate dimensions were already being bandied about in scientific circles.

There are two different kinds of wormholes:

1. Intra-universal: connects one part of Universe A with another part of Universe A.
2. Inter-universal: connects one part of Universe A with a part of Universe B.

Wormholes as Modes of Travel

Some wormholes can connect two different points of time within a universe, and in String Theory, a wormhole could connect two different D-branes. The most exciting kind of wormholes are considered "traversable," meaning they may allow for humans (or non-humans!) to travel through them. Known as Lorentzian Wormholes, these shortcuts through space and time are still theoretical in nature, but some physicists are envisioning ways of overcoming the problems usually associated with the necessity of exotic matter as a requirement for the wormhole to exist.

Traveling through a wormhole would not, as many people might think, mean going faster than the speed of light. In fact, wormholes don't require light speed at all. It is the time that it takes to travel through the hole that gives the impression of faster-than-light travel. All you really need is a black hole and a white hole, and you have your own personal Stargate. The black hole serves as the entry point, and the white hole, the exit. In fact, a black hole *implosion* in one universe means that somewhere there is a white hole *explosion* that creates a new universe.

Okay, so it isn't that easy. In order for a wormhole to be crossed, you first need for the two holes to be stable, and nobody knows what happens to matter once it enters a black hole. Beyond the point of singularity, which you would need to figure out how to survive without gravity stretching you into a super long piece of angel hair pasta, you would need to get through to the mouth of the connecting singularity and out the other hole in what many physicists believe would be a fraction of a fraction of a second before the connection pinches off and collapses into a black hole. Once inside, the tidal forces would be infinite, and you and your atoms would be instantly finite.

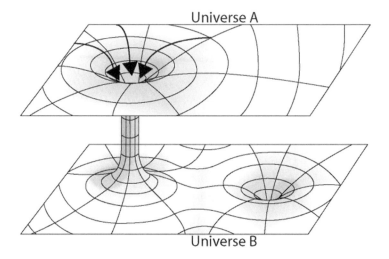

An inter-universal wormhole connects two different universes together.

If you did find a black hole with wormhole potential, it would have to be unbelievably massive, or you would have to be unbelievably small, because many of the wormholes that do exist might be tiny in size, frothing and bubbling up from the quantum foamy sea, only to pinch off and out in an instant.

Then there's the question of just how much radiation you can be exposed to as you enter the black hole, because at the event horizon, all of the light is captured in a big burst of skin-frying fury.

So, wormholes are theoretical in existence, and even more theoretical in just how they might work and what they might do to matter on the way in and on the way out. But some physicists are beginning to wonder if these bridges between space and space, and time and time, might just be travel-worthy after all, at least under the right circumstances.

A One-Way Ticket Through the Universe

Einstein believed that no living thing could ever pass through a wormhole, but in 1963, mathematician Roy Kerr came up with a solution called a spinning black hole that could remain stable because of its outward-pushing centrifugal force canceling the inward force of gravity. Thus, in a Kerr black hole, you would be sucked in and shot through to the other side without being turned into a 20-mile-long piece of spaghettini.

Just one problem. Going through a Kerr black hole means you cannot go home again. You could get through the event horizon in one piece, but there quite simply wouldn't be enough gravity to get you back out again, at least not without another "magic Kerr ring," connecting the universe you entered right back into the one you left.

Michio Kaku, in *Parallel Worlds*, likens it to an "elevator inside a skyscraper." He states that a Kerr black hole would be akin to riding the elevator, connecting you to different floors, each of which is a different universe. You could pick from an infinite number of these floors/universes. But the elevator has no "down" button, so you could never turn back once you choose a floor and exited the elevator.

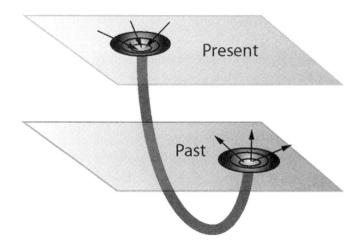

A time wormhole connects two different points in time.

But even Kaku points out that the Kerr black hole, although the closest possible candidate for a traversable wormhole, has its problems, especially with stability. Some physicists believe that traveling through a Kerr ring would contribute to destabilizing it, and that even a beam of light could destroy the "gateway" by generating an interfering gravitational field.

If physicists can figure out how to stabilize a wormhole, they very well could create a time machine, able to send travelers to different points in the past and future, as well as to different places. This would involve sending one end of the wormhole into a future point via rapid motion or gravity. This would also involve massive effort and technology beyond our capability, but theoretically it is possible.

Time Twins and Grandpas

There are paradoxes of time travel that, until solved, continue to puzzle physicists and imaginative science fiction authors alike. One of these paradoxes involves twins. The other, your grandpa.

The Twin Paradox uses relativity theory to show that two twins would age at different rates, depending on their travel choices. Sometimes called the "clock paradox," the TP proposes that if one of two twins journeys into deep space in a rocket traveling near the speed of light, while the other twin remains on Earth, when the space traveler returns to Earth, he will be younger than the twin who stayed grounded.

The Grandfather Paradox deals with causality, namely whether or not you could travel into the past, murder your own grandfather, and live to tell about it. Because if you did kill your own grandfather, you would never exist. Physicists have several ways around this paradox. One is the many-worlds theory, which allows you to go back, kill your grandpa, and continue to exist in a parallel universe that splits off where your grandpa, and thus, you, survived. Stephen Hawking proposed another solution—the chronology protection conjecture, which states that the known laws of physics prevent time travel by "macroscopic" objects such as people. A third option is the Novikov consistency conjecture, which states that physics must not allow paradoxes, so you could go back in time and meet your grandpa for a poker game, but not end his life.

Could There Be Others Out There?

Alice's trip down the rabbit hole serves as a model of what might be expected were we to journey through a wormhole. We might enter another universe or dimension much like our own, or we might enter a totally unimaginable realm where all known laws of physics and common sense vanish the instant we tumble out of the white hole. We might also go to a future point in our own universe, and see the end results of our actions today.

No matter our destination, we would need some way of physically falling down that hole in the first place, a mode of propulsion that could get us across the cosmic freeway and back again. We haven't found that propulsion method yet, but what if other, far more advanced "alien" civilizations out there in our universe, or in one parallel to ours, have? What might they be using for fuel to get here from there?

We know from declassified documents (discussed in Chapter 1), that unidentified flying objects were, and still are, taken very seriously by the governments of many nations. Whether or not these objects came from here, or there, was overshadowed by the desire to know how they worked, and what they might offer in the way of future technology, especially where the military was concerned. If the objects were originating in other countries, the countries that didn't have the technology would do anything to get it.

But a number of UFO cases suggest these vehicles are not originating in some top-secret underground R&D facility, but on some other planet, in some other universe, or possibly some other dimension. The sheer magnitude of the needed technology to traverse across the galaxy (if they are from our universe), or through another dimension (if they are not), suggests that whatever these craft are, they represent a mode of transportation we can only speculate about. The intriguing thing about figuring out how a UFO might operate is this: If they can do it, maybe someday, so can we.

Many of today's top physicists are also suggesting that alien civilizations may be more feasible, thanks to Superstring Theory, parallel universes, and the general acceptance of multiple dimensions, as well as Inflation Theory, which suggests a stretching of space/time. In a January 2005 article on Space.com physicists Bernard Haisch, James Deardorff,

Bruce Maccabee, and Hal Puthoff used Inflation Theory as the springboard for potential ET visitors, suggesting that "we should find ourselves in the midst of one or more extraterrestrial civilizations" and that other dimensions imply the possibility of inhabitable universes right next to ours. They even speculate "it might be possible to get around the speed of light limit by moving in and out of these dimensions."

Types of Civilizations

Michio Kaku, in an article for ABC News titled "Aliens Can Visit, How Did They Get Here?" says "The fundamental mistake people make when thinking about extraterrestrial intelligence is to assume that they're just like us except for a few hundred years more advanced. I say open your mind, open your consciousness to the possibility that they are a million years ahead." Kaku points to the civilization ranking system of Russian physicist Nikolai Kardashev, introduced in the 1960s for the purpose of classifying radio signals from potential extraterrestrial civilizations. Type I, II, and III Civilizations are classified according to their consumption of energy and the laws of thermodynamics. For example, a Type I Civilization has achieved harnessing "planetary forms of energy" and could possibly have mastered controlling weather and building cities upon the oceans. Kaku calls these civilizations "masters of their planet."

A Type II Civilization has mastered the power of a single solar system and colonized much of its home galaxy. A Type III Civilization has mastered the power of billions of star systems. The gaps between civilization types are massive, but Kaku and others believe that they can estimate the time it takes to achieve each type. Take our Earth, for instance. Kaku states that with the trend of economic growth being directly related to energy consumption, we are looking at:

- 100 to 200 years to attaining Type I Civilization status
- 1,000 to 5,000 years to Type II status
- 100,000 to 1,000,000 years to Type III status

That's if we last even 100 years beyond our current state! Carl Sagan believes the gradations should be a bit finer, and places Earth at Type 0.7, a thousand times smaller than a Type I, but even so, we can only imagine the technological breakthroughs that might occur over such amazing spans of time. A Type III civilization would be so far advanced that their technology would be unrecognizable to us. Imagine how someone from the year 1492 might react upon seeing a television, or a rocket soaring into space, or a teenager bouncing around with an iPod and cell phone. They would not believe their eyes and most likely turn away in sheer terror, or utter denial, as many scientists do when faced with the UFO enigma. It would look like magic, or something "paranormal." Insert the *Twilight Zone* theme here!

Kaku points to the Internet as an emerging Type I communications system, and although he feels our civilization is still "primitive" in comparison to Type II, he sees plenty of signs that we are on the verge of a coming transition. It is this gap in achievement and progress, though, that Kaku feels is the reason why we don't see extraterrestrial visitors hanging around our neck of the galactic woods. "If indeed they exist, perhaps they are so advanced that they see little interest in our primitive Type 0.7 society."

What Do They Look Like?

We leave it to science fiction writers to speculate on what intelligent aliens would look like, which would of course be dependent upon the environmental conditions of their habitats. Sir Martin Reese, Astronomer Royal of Great Britain, points out in *Our Cosmic Habitat* that intelligent life elsewhere could be uncommunicative, and that the only type of intelligence we would be able to detect is one that "led to a technology we can recognize." We can only speculate what their ultra-futuristic technology might look like, but in the meantime, we might already have inklings of ways that inter-universal and intra-universal travel might be achieved, if those Type II and III aliens cared to check us out. For decades, scientists have blindly followed the law of science that stated nothing could travel faster than 186,000 miles per second. But what if the speed of light itself is not a constant, opening the door to going beyond faster than light (FTL)? That's what University of Toronto's physics professor John Moffat theorizes. He and fellow U of T researcher Michael Clayton wrote a paper that suggests light once traveled much faster than it does now. During the initial stages of our universe's development, its edges were so much farther apart than light moving at a constant rate could possibly keep up with. Moffat's early calculations, according to Nicolle Walle's article in the March 2003 issue of *SpaceDaily.com*, suggest that light, right after the Big Bang, could have been traveling 1,030 times faster than it does now. Other supporters of his idea that light speed is not constant include theoretical astrophysicist Joao Maguelijo of Imperial College in London, and astrobiologist Paul Davies, who have expanded upon Moffat's work.

The Arrival of Our Cosmic Neighbors

With or without limitations on light speed, and in terms of what is conceivable to our limited minds and imaginations (compared to those of aliens 2 million years ahead of us on the achievement scale) let's speculate on some of the methods "they" might use to pay "us" a visit.

The UFO sightings we are interested in share common characteristics:

1. They leave behind signatures of intense electromagnetic radiation or a strong magnetic field that affects vehicles, energy sources, people, and the environment.
2. They use some method of propulsion that allows them to achieve high speed and aeronautical maneuverability that defies gravity.
3. They operate silently or almost completely silent (as compared to jet engines).
4. They are often associated with a plasma-like corona that suggests some type of electromagnetic or microwave radiation.
5. They are mainly disc, triangular, or cigar-shaped and often display a wobble similar to a spinning top.
6. They accelerate at tremendous speeds and often come to a complete halt, with hovering capability.

These characteristics obviously suggest that these craft are utilizing the electromagnetic field, or at the very least, manipulating it, and possibly generating a field of their own, a field that allows them to do things involving gravity and space/time that make our most cutting-edge military aircraft and rockets pale in comparison. We also know that when our own conventional planes and cars come within close contact of highly intense electromagnetic and/or gravitational fields, their electronics and computer systems go haywire, which would explain many of the "symptoms" reported both in UFO sightings and encounters in the Bermuda Triangle.

Antigravity and Other Theories

We know that research involving antigravity, as well as the control and manipulation of the gravitational field, has been going on for decades. In the late 1950s, American inventor and physicist T. Townsend Brown invented an antigravity rig consisting of a large metal sphere attached via a rod to a smaller sphere. Streaming electrons off a focus-rod electrode attached to a conical electrode charged the larger sphere. The smaller sphere would be charged to the same voltage-pressure as the larger sphere, enhancing the electrogravitic propulsion that would then enable the rig to hover 6 feet above ground. Brown patented two similar objects that worked off of "electrokinetic" propulsion and attraction. His work, after the Second World War, would be continued as classified research at a number of major multinational corporations such as General Electric, Convair, and Lear.

Brown's rig sounds a little bit like the prototypes made by German engineer Viktor Schauberger, as discussed in Chapter 8, who designed various objects that worked off vortex energy, and possibly Zero Point Energy. Schauberger's levitating discs used an implosion method of three-dimensional spiraling energy patterns channeled inward, rather than outward, creating massive levels of force.

Later, in the mid-1990s, Russian physicist and gravity researcher Evgeny Podkletnov would engineer a device that shielded gravity by 5 percent (some critics argue only 2 percent), which he would present to NASA's Breakthrough Propulsion Physics program. This 12-inch diameter ceramic donut-shaped superconductor ring was able to levitate and spin when placed over solenoids. The outer steel casing of the ring contained liquid nitrogen, which cooled the device as it spun. This device was made to spin at 5,000 rpm. Objects placed over the spinning ring would lose a percentage of their weight. Podkletnov also discovered a circular funnel of lesser gravity 1 foot in diameter above and below the device.

Podkletnov would also announce during a 2004 interview his creation of a "gravity-beam" that uses a 5-million volt discharge with peak-power density in the terawatt-range. This experimental beam is supposedly capable of exerting hundreds of pounds of force onto an object, thus able to penetrate solid materials without a loss of energy. Podkletnov's gravity beam could lead to breakthroughs within the next decade in applications for both the transportation and defense industries.

From Tim Ventura's levitating triangular lifters to Hutchinson's Effects that cause objects to levitate and fly across the room, the ability to toy with gravity could provide an alien civilization with a potential method for achieving near-light speed travel. Many UFO reports indicate that the craft are able to either create a gravitational field to counter that of Earth's, or control it. The gravitational field could be made to accelerate, with the craft snuggled safely inside, rocketing the aliens to the farthest reaches of space without increasing mass, curving space, or messing with time so much that leaving home would result in having no home to go back to.

In February of 2006, noted physicist Dr. Franklin Felber presented a paper to the Space Technology and Applications International Forum in Albuquerque, New Mexico, on his new theory for enabling space travel near light speed. His research shows that mass moving faster than 57.7 percent light speed gravitationally repels other masses that are lying in the range of a narrow "antigravity beam" and that this beam becomes stronger the closer the mass gets to the speed of light.

Felber theorizes using this "repulsion" of a speeding mass through space as a potential source of energy, enough energy, in fact, to possibly accelerate massive payloads that draw energy from the antigravity beam. And the larger the mass, the larger the beam, and thus, the greater the energy. Felber's theory is just that—theory—but his solution to the light-speed issue could be tested in a lab.

Other theories involving antigravity include building a machine that could create a gravitational field that would then counteract with that of the Earth's own field. Anything within that "zone of counteraction" could literally defy gravity as it traveled along, limited only by the size of the fields themselves. However, generating a large enough force to counteract Earth's gravitational field would require more technology than we have, and more mass than we've got. It's one thing to lift a triangular-shaped lifter made of balsa wood and tin foil, and quite another to create the necessary "lift" to propel a ship out into the far reaches of space.

Zero Point Energy: Fuel for the Gas Tank

Intra-universal and inter-universal wormholes might not require faster than light capability, but would still require vast energy sources. The Zero Point Field is one potential source of unlimited, "free" energy that a space vehicle could tap into. Because Zero Point Energy is present in every corner and crevice of space, it would not limit travel to any particular part of the universe.

ZPE spacecraft could solve the two main space travel problems of speed and fuel supply. The quantum fluctuations of ZPE must first be extracted from the vacuum, and engineering a machine effective enough to do so is the only thing many physicists believe stands in the way of human space travel beyond our wildest dreams. It is certainly feasible that advanced alien technology has found a way to extract the ZPE, and according to the documented research of Nazi interest in the subject matter, as pointed out in Chapter 8, we may be able to duplicate it. Delving deeper into the Casimir Effect might hold the key. We know from the Effect that the energy between two metal plates held at close distances is less than the energy outside of the plates, which pushes the plates together. The same goes for the radiation pressure, which leads to the conversion of vacuum energy to heat between the plates.

So far, this and other experiments with extracting ZPE have led to extremely small doses of energy, but ongoing research into engineering better extraction methods will no doubt change the future of not only space travel, but fuel sources for more domestic, Earthbound vehicles.

ZPE and antigravity both offer tantalizing methods of UFO propulsion, but also offer cheaper fuel sources for a world "addicted to oil." Some researchers, such as Jim Marrs, author of *Alien Agenda* and *Rule By Secrecy*, suggest that these alternative technologies lack the funding and attention they require because of the monopolies of interest, such as oil and gas companies, that would suffer from the discovery of such abundant and cheap fuel sources being made available to the public. (Marrs is also quick to note that antigravity and the ZPF might explain the many UFO sightings involving automobile engines stopping and starting up again, and other such examples of the manipulation of energy reported.) Hal Puthoff, in an interview with *Fortean Times*, also speculated that someday the ZPE could be used in the process of water desalinization, as well as reducing dependency on fossil fuels. It may also play a role in heating homes, and if a process is found to convert this energy into electrical form, Puthoff suggests we may one day have batteries that far outlast the Energizer Bunny!

The great thing about using ZPE as a fuel source for skipping across space is that your ship can easily "stop for gas" along the way to another universe, and the ship itself would be free from carrying a weighty amount of fuel to slow it down. The ZPE exists in a virtually infinite reservoir from which power can be obtained. A civilization advanced enough to figure out the engineering solutions could pretty much go anywhere it wants, and in record time. Puthoff pointed out to *Fortean Times* that a UFO could possibly use ZPE as either a fuel source, or as a means for "the perturbation of the space/time metric," and suggested ZPE might also account for natural atmospheric anomalies that some people mistake for UFOs, such as ball lightning.

Not all physicists are jumping on the ZPE bandwagon as a potential source of fuel, or even as a form of energy in empty space. Professor Steven Weinberg, in an interview with *Scientific American*'s "Ask the Scientist" series, referred to the law of conservation of energy, which tells us that if we get energy out of empty space, we must leave it in a condition of lower energy. What, he asks, could have lower energy than empty space? But Puthoff cites modern research that shows a vacuum state such as the ZPF can have different energy values, and can even decay to a state of lower energy under specific conditions. In the same interview series, Puthoff mentions laboratory experimentation into ZPE that could yield extraction methods. Some of these experiments have shown results on a small scale, but with further research and funding Puthoff hopes that within the next decade, "we will either be confident that it is only a matter of time and engineering, or it will reveal itself to be only a laboratory phenomenon without the possibility of constituting a major energy resource." He compares attempts to harness ZPE to a "long list of harnessing energetic processes we find in our natural environment."

The Zero Point Field and its unlimited energy just might be the fuel of choice, then, for traveling aliens with highly advanced technology eager to take a peek at Earth. But it may not be the only choice.

Other Matters

It's a given that our chemical rocket propulsion methods can't provide enough energy to get a vehicle across space, let alone into a wormhole and out the other side. But there is an explosive method of propulsion that alien civilizations could be using, one that we humans are just beginning to understand, and dare to work with.

Antimatter, similar to antigravity, involves the annihilation or control of its opposite, in this case, matter. Matter and antimatter have opposite charges. By combining matter and antimatter, and theoretically any element can have an anti-element, the end result is a tremendous amount of energy. The mutual annihilation that occurs when matter and antimatter meet converts all their combined mass into energy, but it also produces high-energy radiation that would need to be dealt with in terms of protecting a potential crew. Another downside is the lack of natural sources of antimatter, and the difficulties involved with creating enough to fuel a rocket ship to the nearest star. However, if an alien craft had the capability of tapping an unlimited source of antimatter already out in deep space (or another universe) or the ability of converting massive volumes of matter into antimatter quickly and easily, then they indeed could get here from there . . . and without frying its crew members in the process.

Negative matter is another potential method of star flight propulsion. Not to be confused with antimatter, negative matter has negative mass, not a reverse electrical charge from matter. Negative matter could have a reversed mass polarity that could possibly nullify the mass of its positive matter. This "charge imbalance" between positive and negative matter could accelerate a vehicle through space, according to a concept proposed by astrophysicist Hermann Bondi way back in 1957. The only problem is, nobody has ever found a trace of negative matter.

Speaking of negative things, physicist Kip Thorne postulated in 1988 that a wormhole could become traversable if a small amount of exotic matter with "negative energy" could be put into the throat, or tunnel. This exotic matter would literally have less energy than ZPE, thus making it negative energy, and some physicists believe only a tiny amount of exotic matter would be needed to make the wormhole passable.

In principle, threading its throat with exotic matter, which has negative mass and positive surface pressure, could stabilize a wormhole. The negative mass ensures that the wormhole's throat lies outside the horizon, allowing spacecraft to go through it while the positive surface pressure stops the wormhole from collapsing. As long as the exotic matter is contained along the edges of the frame that supports the wormhole, its mouth can be open for business—the travel business. (Count me out, I get motion-sick watching the Travel Channel!)

Thorne, in fact, devised a wormhole concept that did not require a black hole and a white hole or that nasty event horizon with its point of singularity. He proposed that the throats, or tunnels, could be large enough for a human in a craft to get through, and that there would be no threat of the mouths pinching off. Instead, his wormhole would be stable and remain open for two-way travel, and would link two points in two-dimensional space. But Thorne admits this traversable wormhole most likely does not exist naturally, and that having the technology to actually build one of these is light-years beyond our ability.

Time Travel

Time travel enters the picture here, because any advanced civilization will have no doubt learned to control time as well as space. Wormholes offer the opportunity for moving between parallel universes and extra dimensions, but also for moving between past, present, and future. The idea of a time machine is nothing new, having been a staple of science fiction and fantasy novels and movies throughout the last century. But few people are aware that we humans have succeeded at building time machines, as documented in Jenny Randles' *Breaking the Time Barrier: The Race to Build the First Time Machine*. Randles documents the research into time travel from the time of Marconi and Edison, straight through to the most cutting-edge lab experiments being conducted today. Amateurs and professional scientists alike seem bent on bending time to our bidding, and according to Randles, some may indeed have time on their side.

Most of the researchers Randles documents involve building a machine for human use, such as the Tipler Cylinder. Proposed in 1974 by Tulane University's Frank Tipler, this intriguing machine would involve a rotating cylinder made of dense matter and spinning thousands of times per second. Basically, you put that cylinder out in space and travel in a spaceship to the cylinder, get close to the surface of the spinning cylinder, orbit around it a few times, and then go home. When you got back home, you would be in the past, depending on how many orbits you made. That's because the warped region of space around the spinning cylinder acts as a time suppressor, moving anything outside the warped region further and further into the past.

Tipler's time machine required that the cylinder be infinitely long, but other researchers believe the same concept might work without the problem of the infinite. In fact, some suggest you could travel through time by finding a spinning black hole or neutron star to orbit around. Another idea involves infinitely long and dense cosmic strings that could be made to spin fast enough that a closed time loop would form around it. You wouldn't even need the string to spin, according to Princeton University theorist Richard Gott, who in 1991 suggested that all you would need to create a time loop are two strings moving past one another at high speed. They would, however, have to be at just the right angle, parallel to each other as they passed. First, physicists need to prove that cosmic strings exist. Gott and his colleague Li-Xin even devised a bizarre cosmological model of an entire universe tracing a loop in time so that its end would be its beginning. (Where's that bottle of aspirin?)

Randles writes about physicist Ron Mallett, who set about building a time machine for personal reasons. Mallett was only 10 years old when his father died, and he felt cheated out of knowing his dad; thus, the quest to construct a time machine and possibly see his father again. Basically, Mallett's machine is designed to warp space using lasers placed in orbit and rotated at high speeds to create a vicinity of distorted time within. He believes that in this zone, time might take on the effects of a spatial dimension, a time with its own landscape, that you could move around in.

Whether Mallett's ongoing research into this laser warp time-drive, which began in earnest in 2002, results in a real and workable time machine, only time itself will prove, but according to Randles, even if he only succeeds at controlling the flow of particles, it opens the door to conveying "meaningful information through the time barrier." Mallett, an advocate of parallel universe theory, was featured in an interview with Physorg.com, a Web site devoted to science and technology, where he compared his experiment to a sugar cube dropped into a cup of coffee. "The coffee is empty space, and the spoon is the circulating light beam. When you stir the coffee with the spoon, the coffee—or empty space—gets twisted." According to Einstein, if you do something to affect space, it would also affect time.

Mallett thinks his time machine could also solve the Grandfather Paradox, due to the presence of other universes. A time traveler, he states, would not just travel through time, but through other universes.

Time Storms and Other Mysterious Occurrences

Human-engineered time machines, wormholes, and other modes of moving between universes aside, alien technology may be way ahead of us in terms of engineering the energy of the cosmos for practical uses. But Randles points out in another of her books, *Time Storms*, that time warps (and the inherent time travel potential they contain), may already exist naturally. Her book documents dozens of cases of people who have experienced missing time, time leaps, warps in the perception of time, and even rifts in space itself. In terms of UFOs, existing time warps may come in the form of energy vortices, such as the Bermuda Triangle, the Great Lakes Triangle, the 12 vile vortices, and other hot spots around the world where ships and planes disappear, and often objects of unknown origin come visiting in their place.

Time storms can be either stationary, as in the Triangle, or moving, as in the many fog-like masses reported in Randles' book. People entering these mysterious, and often traveling, fogs encountered electromagnetic anomalies and experienced gaps in time, sometimes even finding themselves teleported over vast distances, with no recollection as to how they got there. If we have these mysterious cases of time travelers right here on Earth, perhaps UFOs and other strange creatures associated with hot spots may be coming to our neck of the woods through the very same energy portals. After all, what goes up must come down, and what goes in must come out somewhere.

Could disappearances of planes and ships, such as those documented by the U.S. Coast Guard and paranormal investigator Gian Quasar, author of *Into the Bermuda Triangle*, simply be entering such time storms or "slips," only to come out in some other universe or dimension, just as objects we cannot identify might be coming here? If indeed places such as the Triangle are Earthbound wormholes connecting two points in our universe, two points in different universes, or two points in time, it would serve to explain both the hundreds of vanishings on our end, and the thousands of appearances on our end.

The electromagnetic effects often reported both in UFO sightings and in time storm experiences suggest that something is happening to alter the usual space/time fabric, and allow a rip or hole to occur. They come here, we go there, and sometimes we come back. Sometimes, as in the case of the crew of Flight 19, we don't.

Randles is convinced that time storms are perfectly natural occurrences, and paranormal investigators are convinced that the Bermuda Triangle may just be one big, fixed time storm, but one that fluctuates in intensity, which would explain why some planes and ships make it safely through while others do not. Gian Quasar compares the antigravity device discovered by Podkletnov, described earlier, to the naturally occurring "funnels" discovered by Canadian Wilbur Smith. These funnels, described in Chapter 3, extend above the ocean to high altitudes and contain magnetic and gravitational anomalies. Smith believes the Bermuda Triangle is one such area where these mysterious funnels exist, and that their intensity could vary, thus resulting in planes and ships vanishing into them one day, but not another, depending on various intensifying factors affecting these magnetic vortices. Again, we are reminded of Ivan Sanderson's "vile vortices" operating all over the globe at precise 72-degree intervals.

Quasar also refers to the Hutchinson Effect as a possible theory behind the Bermuda Triangle, citing the ability of objects to levitate, appear and vanish, and, in the case of water, swirl and create mile vortex kinesis. Maybe these combined electromagnetic-gravitational effects operate in the Triangle on a massive scale as compared to Hutchinson's home lab, and could result in enough spiral movement to send a plane or a ship upward into another dimension, or universe. Quasar even reports Hutchinson himself as saying the Bermuda Triangle might indeed be related to his discoveries. He told Quasar, "It is highly probable that nature can form these fields on her own and create the right situation for the ship or aircraft to either totally disintegrate or disappear into another dimension or domain."

A team of Brown University scientists has already managed to levitate the embryo of a frog using a technique called "magnetic field gradient levitation," which puts an object in near zero-gravity conditions. This mirrors the connection between PK and poltergeist effects such as levitating objects and moving furniture and strong magnetic fields present at the time. (Not just in nature, but in the individual's brain!) Magnify that magnetic field gradient and who knows what you could lift . . . or what could lift you?

If physicists ever do prove the existence of wormholes and time warps, they may want to look for those right here on Earth first. Naturally occurring magnetic vortices of energy hold many clues to how gravity can be altered, and how the warping of space and time is possible under just the right electromagnetic conditions.

Most scientists still refuse to believe anything that smacks of the paranormal, and unfortunately, for many, time travel falls into that category. They suggest that if time travel were possible, we would be seeing a lot of time travelers. They conveniently ignore, simply because they cannot explain, the thousands of UFO cases that suggest time travelers have been visiting us all along. Nor do these scientists pay much attention to the hundreds of declassified government documents that state objects have been coming into our airspace that cannot be categorized as anything else but "out of this world."

Playing With Time

Perhaps, because of the theory of parallel universes, time travelers from distant places just don't feel much like coming to a planet similar to ours. We are, perhaps to them, gaudy.

But those who are open to the possibility of time travel do argue that there are real, physical limitations to consider. It's the vague and arrogant limitations, such as the suggestion that time travel to the past is impossible because the current known laws of physics forbid it, that keep real progress stuck in slow-mo.

UFOs can get here from there, if they operate under their own set of physical laws, or if they, unlike us, have learned to master the laws we operate under. Truly, the majority of UFO sightings can be proven as something other than alien craft from another world, but for the many that cannot, why on Earth, or beyond, would we limit their capabilities to our own? They apparently manipulate the electromagnetic field, toy with gravity, and utilize an energy source that allows them to travel at speeds and distances beyond human capability.

The exciting thing is that manipulating the EMF, toying with gravity, and tapping the ZPF are all within our reach—maybe not today, but who knows what dreams may come?

Ghosts and Time Travel

Warps, fields, and wormholes may provide the means of travel for alien civilizations, but they may also provide other phenomena with the means of entering our world from another dimension, another place . . . another time. Paranormal investigator Joshua P. Warren thinks ghosts may also use such portals, sighting that certain locations seem to attract greater paranormal activity than others. If the Earth's magnetic field is not balanced, because the Earth's physical form itself is not perfectly balanced, then Warren believes there may be areas of unusual geomagnetic activity that could cause reality to behave differently.

Warren calls them "warps," and suggests that ghostly effects may result from distortions in the laws of physics. He cites his own research into a variety of ghostly manifestations associated with electromagnetic disturbances, as well as poltergeist activity. Warps, in terms of ghostly phenomena, are areas where linear time doesn't apply, perceptions can be twisted beyond logical understanding, and where entities and other paranormal activity exist.

When we think of ghosts, we think of the past. Ghosts are usually the entities, specters, or images of a person or thing that existed in the past, and are now appearing to us in the present, which to them is the future. This implies time travel on the part of the entities, and an ability to cross through from one dimension or point in space/time to another, quite like the UFOs we discussed above. Warren refers to some ghosts as "imprints." These are ghosts that do not appear to be conscious and are oblivious to observers. Imprints often repeat the same actions over and over again, as if they are the "residue" of a memory that has been impressed upon the environment.

This supposed "recording" of the energy of an event could take place in the Zero Point Field, which we earlier compared to the Akashic Records of Edgar Cayce, upon which every memory, action, thought, and thing was written. These imprints, or recordings, could have found a way to exist intact upon the ZPF, and those who see ghosts could have found a way to tap into them. More on that in the next chapter.

Traveling Isn't Just for Humans

Some cryptozoologists believe mythical creatures such as the Mothman and the Loch Ness Monster may be inter-dimensional travelers, also able to move between worlds via warps or portholes right here on Earth (or beneath the water). Many of the sightings of strange creatures documented by investigators such as Nick Redfern and "monster-hunters" Jon Downes and Richard Freeman in *Three Men Seeking Monsters* suggest the ability to appear and vanish instantaneously (similar to some UFOs), tamper with electromagnetic fields (similar to some UFOs), and even emit electrical static and discharge (similar to some UFOs). These creatures also seem to exist in particular places, places that might be home to some of the phenomena also associated with time storms, such as mysterious fogs and higher than average EMFs. In the case of poltergeists, William Roll discovered that many "events" occur in conjunction with solar flares, magnetic storms, and increased sun spot activity, which affect the Earth's electromagnetic field and cause all sorts of crazy things to occur, from garage doors opening and closing on their own, to electrical failures. Roll cites the work of Michael Persinger, the renowned neuropsychologist who found that geomagnetic disturbances interfered with ESP.

Persinger's research proved a significant correlation between such geomagnetic disturbances and the increased ability of recurrent spontaneous psychokinesis (RSPK), discussed in detail in Chapter 4. These disturbances affect the brain and have even been shown to bring on epileptic attacks, according to research documented in medical journals.

Roll found the connection between geomagnetic eruptions and RSPK activity logical, because RSPK was initiated in the brain by a similar process to epilepsy. Because epilepsy was affected by environmental conditions such as increased magnetic activity, so then must RSPK be duly affected?

The English countryside has more than its share of cryptozoological enigmas running through field and forest, just as the Bermuda Triangle has more than its share of downed planes and compass and electronics malfunctions. Might it have something to do with the location of vile vortices, ley lines, or electromagnetic "burps" (if you will)? Higher bursts of sunspot activity, geomagnetic storms, or alterations in the normal EMF ranges? Literal "zones of enigma" that contain all the right electromagnetic elements necessary for warping space, perturbing time, and distorting the lines between our world and theirs?

In other words, could aliens, beasties, and ghosts be using wormholes and the Zero Point Field to get here from there? Possibly, but there are other ways that paranormal activity might manifest in our world, no matter where it comes from . . . or when.

Resonance:
Synchronized Swimming in the Quantum Sea

Resonance (n): the effect produced when the natural vibration frequency of a body is greatly amplified by reinforcing vibrations at the same or nearly the same frequency from another body.
                                        —Webster's New World Dictionary

I am convinced that there are universal currents of Divine Thought vibrating the ether everywhere and that any who can feel these vibrations is inspired.
                                        —Richard Wagner

Things we do and experience have resonance. It can die away quickly or last a long time; it can have a clear center frequency or a wide bandwidth; be loud, soft or ambiguous. The present is filled with past experience ringing in various ways and now is colored by this symphony of resonance.
                                        —Paul Lansky

As a kid, watching the synchronized swimming competition of the summer Olympic games, I was always struck by how silly the sport seemed. Not to mention how it looked! Later, as a wise adult, I came to understand and appreciate both the degree of difficulty of getting human bodies to move in perfect unison, and the beauty that results when they do.

But bodies are not the only things that resonate. We know from past chapters that there is no such thing as empty space, and that the so-called vacuum of space is teeming with quantum fluctuations that display a resonance, a vibration. Nothing does not vibrate. There is no such thing as zero, dead, still. Everything that exists gives off some vibration of a certain frequency. Every planet, every person, every particle.

Musicians know that when a guitar string is plucked on one instrument in a music store, all the other guitars in the same room will vibrate to that tone. Healers refer to this as "entrainment," when two objects (or people!) in close proximity, vibrating on different frequencies, begin altering their vibrations until they are vibrating at the same, or nearly the same, frequency. The Zero Point Field could act as a field of "entrainment" or resonance, where the vibrations of particles tune to specific frequencies, creating different forms of matter, energy, and interactions. Other scientists believe this field has different names.

Quintessence

In 1998, physicist Paul Steinhardt and his colleagues coined a term that would describe a mysterious field of what they believed to be dark energy, or a "fifth essence." They called in "quintessence." Based upon an earlier idea proposed by Fermilab physicist Chris Hill and colleague Josh Freeman, quintessence suggested that, like the cosmological constant of Einstein's vision, this essence fills all of "empty" space with a form of matter-energy that is changeable in strength. Some parts of space might have a thicker quintessence, others a thin "layer," creating a field of invisible essence that has no direction, such as a vector field, only magnitude, as in a scalar field.

Described as kinetic energy by Tom Siegfried in *Strange Matters*, the strength of this field would be measured by how quickly it approaches the zero point. Because quintessence is believed to exert negative pressure, it is said to be a slow rolling scalar field, one that does not change too quickly over a period of time. Whether or not quintessence is the dark energy so eagerly sought by physicists will decide the fate of the universe itself, because of its relationship to expansion. The presence of a negative energy in space would possibly stop the expansion. And, if dark energy were indeed the same as quintessence, the changeability of the field would suggest the universe's fate is one we cannot predict with accuracy.

Aside from the role quintessence plays in the outcome of our universe's destiny, it may also play a role in the way matter and energy interacts, or resonates. Nature is filled with signs of the importance of resonance and the beauty of synchronicity. From the physical foundation of all musical composition to the intricate mathematical ratios of the natural world, there seems to be an element of "arrangement" that results in a visible pattern.

Patterns in Reality

Take the Golden Ratio, known since the dawn of Egyptian civilization, which is found in art, architecture, and the human body. The Golden Ratio is a representation of the ratio 1.618, otherwise known as phi. This measurement of one-half the sum of 1 and the square root of 5, has been used in famous paintings as well as buildings such as the pyramids to create a perfect ratio that mimics those found in nature.

The Golden Ratio, also referred to as the Golden Section, Golden Mean, divine proportion, or "sectio divina," is found in the form of another phenomenon known as the Fibonacci pattern, or spiral. Look at the positions of leaves on a flower stalk, or the spiral shells of

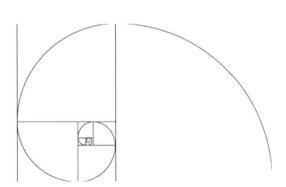

The Fibonacci spiral is found everywhere in the natural world.

mollusks, or the positioning of seeds in a sunflower's face and you see evidence of a Fibonacci spiral.

The spiral is represented by a series of numbers discovered in the 12th century by Leonardo Pisano Fibonacci, a mathematician who first associated the natural patterns evident in nature. The numbers form a series that, starting with the third number, are the sum of the preceding two numbers: 1, 1, 2, 3, 5, 8, 13, 21, 34 . . . . As you go farther and farther down the Fibonacci sequence, the ratio of two successive Fibonacci numbers oscillates about (being alternately greater or smaller) but comes closer and closer to the magical number—the Golden Ratio, an irrational number (never-ending, never-repeating) with an approximate value of 1.618. Otherwise known as phi.

You Are Golden

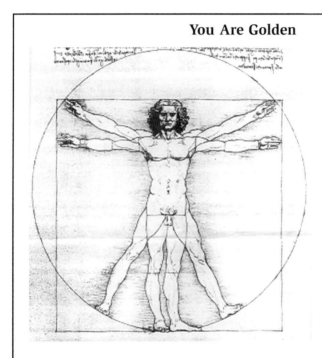

The Vitruvian Man, made famous by Leonardo Da Vinci (which has been "reborn" thanks to the popular novel, *The Da Vinci Code*) shows the Golden Ratio applied to the proportions of the human body. The average human body sports a ratio of 1.618 when you measure your height in centimeters and divide it into two measures. The first measures the distance from your feet to your navel (Long Measure). The second measures the distance from your navel to the top of your head (Short Measure). The first example of the Golden Ratio states that if the Long Measure (navel to foot) is defined as 1 unit, then the height of the entire body is approximately 1.618 units. The second Golden Ratio compares the Long Measure to the Short Measure. Some of the other golden proportions of the human body include:

> Distance between the fingertip and the elbow/distance between wrist and elbow.
>
> Distance between shoulder line and top of head/head length.
>
> Distance between navel and knee/distance between knee and end of foot.
>
> There are even Golden Ratios found in the ideal human face, with length of face/width of face; length of mouth/width of nose; and distance between pupils/distance between eyebrows.
>
> Nature is made up of stunningly intricate patterns, and, apparently, so are we. Da Vinci believed the Vitruvian Man, which was completed in 1490, was the perfect artistic example of the "cosmografia del minor mondo"—the cosmography of the microcosm. He believed that the workings of the human body were an analogy for the workings of the universe.
>
> As above, so below . . . As within, so without . . .

These amazing mathematical measurements found in the natural world (and replicated in the human world of art and architecture for aesthetic value), indicate a set of universal laws, or rules, that are at the very heart of creation itself. These laws indicate a profound resonance between energy and matter, resulting in the most intricate specifics of a tropical mollusk's shell or the petals of a rose. Scientists claim the Golden Ratio is even present in snow crystals, viruses, and microscopic organisms, as well as in the behavior of light rays in experiments with layers of glass. If our universe is really finite, with boundaries, as suggested in earlier chapters, and that we are but one universal bubble floating in a sea of bubbles, we might speculate that, were we ever able to measure the boundaries of our universe against those of the larger field of existence, that measurement might be approximately 1.618.

Sacred geometry utilized the patterns of nature and the presence of energy grids when constructing buildings such as churches, temples, and places of worship. The term "sacred geometry" covers both Pythagorean and Neo-Platonic geometry and encompasses religious, philosophical, and spiritual beliefs into the physicality of geometric design. Ancient Greeks especially assigned attributes to Platonic solids and geometrically derived ratios that had a spiritual meaning, and the Golden Ratio or Section was one of those dynamic principles that embodied the physical and spiritual power of patterns and vibrations. In fact, sacred geometry has been said to both tap into, and create, vibratory patterns in the Earth, and in the cosmos, as many sacred sites are often built in alignment with stars and planetary movements. Sacred architecture was built on specific locations, and in specific patterns that were symbolic of geometrical archetypes, which reveal the nature of form, reality, and the underlying vibrational resonance.

The concept of vibrational attraction has been tested in a more scientific setting, including studies by quantum biologist Dr. Vladimir Poponin, who created a vacuum in a container. The only objects left were photons—light particles, which were measured and found to be randomly distributed . . . until Poponin introduced physical DNA into the container. The photons then lined up in orderly fashion, precisely in alignment with the DNA. Once the DNA was removed, the photos *remained in the order of the DNA.*

The end result was surprising, suggesting a resonance between the photons and the DNA, even after they were separated.

If we can speculate that there exists an underlying pattern or structure to reality that manifests on the visible scale, a field of essence, of pure vibrating energy present throughout space and time, then we may also speculate upon the nature of this field, and what might happen when those vibrations occur in a synchronized state, the way guitars in a music shop will hum "in synch" to create a harmonic convergence of sound.

The Implicate Order

Physicist David Bohm believed that underlying the physical, tangible world, there is a far more mysterious, deeper order of "undivided wholeness." He called the visible world the explicate order, and that deeper world the implicate order, and used the analogy of a flowing stream to describe his realization of unbroken unity.

"On this stream, one may see an ever-changing pattern of vortices, ripples, waves, splashes, etc., which evidently have no independent existence as such. Rather, they are abstracted from the flowing movement, arising and vanishing in the total process of the flow. Such transitory subsistence as may be possessed by these abstracted forms implies only a relative independence or autonomy of behavior, rather than absolutely independent existence as ultimate substances."

Those rather philosophical words came from Bohm's *Wholeness and the Implicate Order*, which he wrote in 1980, and suggest the world of the implicate order is similar to a hologram, where the complex interference patterns appear to be chaotic and disordered to the naked eye, yet on a deeper level possess a pattern that is hidden or "enfolded" into the whole object. Bohm even suggests the universe itself is like a flowing hologram, or "holomovement," that contains order on an implicit level. The explicate order would be the projection from higher dimensions of reality, and any apparent stability of objects and entities are really a sustained process of enfoldment and unfoldment. Nothing solid is really solid at the implicate level.

Bohm also believed there is a superquantum potential, or a higher "superimplicate" order operating in the universe, and that life and consciousness, which we will discuss in detail in the next chapter, were enfolded within, with matter appearing in varying degrees of "unfoldment." He even stated that the separation of matter and spirit is nothing but an abstraction, that the "ground is always one."

The implicate order is the fluctuating sea of energy from which matter unfolds alongside it. One is visible, like a television. The other, invisible, such as the electricity that powers the TV. The study of music (and the patterns of sacred geometry) suggests an invisible, "implicate" vibratory nature. Sympathetic Vibratory Physics is a term assigned to a "musical universe," an interesting alternative theory of reality proposed by Walter Russell in *A New Concept of the Universe*. Russell believes that there exists nothing in nature other than vibration. He attempts to create a paradigm of reality using wave and vibration theory based upon the work of John W. Keely's concepts of a sympathetic vibration that connects all things and energies, and that the harmony of these vibrations creates what we see.

Variations of Nature

Russell and Keely suggest that music can be thought of as a model of the order found in the universe, with organized vibration or sound following principles of structure and behavior that make sound into harmonies. These principles mimic those governing other vibratory patterns in the universe.

Take the idea that everything is the result of a vibration, and even Zero Point Energy in the quantum vacuum has been shown to "jiggle" or vibrate. Everything has its "jiggle." We can go on to say that different things have different "chords" or "vibration signatures," and that is what makes one thing discernable from the next. Sounds a lot like Superstring Theory, with tiny, vibrating cosmic strings at the very heart of existence.

Vibrations are dynamic, interacting with one another and their environment, creating different tones and chords and harmonics. In a sense, vibrations are more fundamental to reality than the tiniest subatomic particle, because in a sense, that is what the tiniest particle is . . . a vibration. We can also add to the mix, pun intended, the various pitches and frequencies, as well as musical intervals, and easily see how the universe could be made up of harmonics. After all, isn't it called a "uni-VERSE?"

The most exciting thing about this invisible vibratory field, no matter what form it might take, is the potentiality it contains—potential as a source of all other sources, potential as a pure energy field upon which all matter is created and thrown out into the explicate order, and potential as a doorway to other dimensions or levels of reality, where things such as ghosts, UFOs, and psychic abilities are the norm.

Theories Abound

Physicist Paul LaViolette writes extensively about "transmuting ether" in his book, *Genesis of the Cosmos: The Ancient Science of Continuous Creation*. This ether, which is the basis of LaViolette's theory of "subquantum kinetics," is described as an active substrate that differs from the mechanical ethers once considered in previous centuries. LaViolette states, "The concentrations of the substrate composing this ether are the energy potential fields that form the basis of all matter and energy in our universe." The ether reactions cause wave-like field gradients that emerge and form the "observable quantum level structures and physical phenomena," including the various particles, forces, fields, and electromagnetic waves.

Subquantum kinetics, which LaViolette proposes as a unified field theory that fills in the gaps conventional physics cannot, rests on the existence of this "primordial" transmuting ether present throughout space. "The transmuting ether is the wellspring of Creation," he states, adding that were its activity to diminish, everything physical would cease to exist, coming to a state of "multi-dimensional consciousness" from which the physical universe is generated. A ground state, or Source of all Sources.

Again, this ether sounds a lot like the Zero Point Field.

Rupert Sheldrake, biologist and coauthor of *The Evolutionary Mind: Conversations on Science, Imagination and Spirit* refers to this vibratory field as the "morphogenetic field." This "M-field" (and there can be many M-fields) is an underlying energy field that acts as an organizing principle to give form to various levels of reality. He also suggests there is "morphic resonance," or the resonance of memories that exist in the sort of Akashic field or "collective unconscious" of Jung. These memories shape our minds today, but subconsciously. Sheldrake, who has experimented with psychic pets and animals, theorizes that the existence of an invisible field of influence, although unproven, could be the link between humans and their pets. "Morphic fields also contain attractors, which draw organisms towards future states." This could explain how a dog might pick up a change in the morphic field that lets it know it's owner is only a half block away.

This invisible field could also be the link between humans and their ghostly visitors, and between humans and poltergeists, as discussed in William Roll's *Unleashed*. The presence of both electromagnetic anomalies and the theory of the Zero Point Field both played a part in Roll's investigation, detailed in earlier chapters. Playing off of the magnetic fields, the way birds do when homing, the poltergeist energy found an outlet for manifestation through the brain of a young teenager, and many ghost hunters will point to energy fluctuations and EMF changes present during a haunting or sighting event.

The morphogenetic field itself could be made up of "morphic wavelets" of resonating, vibratory energy that differ in scale and frequency. The wave could have a resonance that synchs up with the resonance of a pet, or a human, who then displays some "psychic" ability to predict a coming earthquake or disaster. And if they are tapping into a field that contains memories of past, present, and future, a time landscape so to speak, they could access any event throughout the space/time continuum.

It sounds spacey, but if we recall that there is no linear time, except in our own brains, and if we recall there are higher dimensions, parallel universes, and fields of vibratory resonance, then it seems quite possible humans and animals could find a way to set their own frequency to the station of their choice.

Obviously, it's not that simple, or we would all be seeing ghosts, aliens, strange creatures, having precognitive dreams, major déjà vu episodes, and reading each other's minds. But if even one small percentage of the claims of the paranormal is true, someone somehow is figuring out just how to do it.

The Theory of Synchronicity

Terence McKenna, psychedelic visionary and coauthor of *The Evolutionary Mind*, points out "Once nonlocality is accepted, some of the things we're interested in are permitted—telepathy, information from other worlds arriving by the morphogenetic field, and so on." We know that experiments have proven nonlocality to be a reality on the quantum level. Two particles continue to affect one another at extreme distances. The same actually happens on a macrocosmic level all the time.

We call it synchronicity.

Advanced physics tells us that an event at Point A in the universe does not cause an event in Point B to occur a little later. Actually, both events occur at the same time. Carl Jung used the term "synchronicity" to describe this same phenomenon on a human level. In his amazing book *Power vs. Force*, Dr. David R. Hawkins states that a "question can't be asked unless there's already the potentiality of the answer...there can be no 'up' without an already existent 'down.'" Synchronicities are evidence of an all-inclusive field that goes against the normal cause-and-effect rationale for events that occur that challenge our illusion of the line between subjective and objective reality.

Perhaps, the universe itself is based upon principles of synchronization. Physicist Claude Swanson, educated at MIT and Princeton University, believes it is. He even believes his "synchronized universe" theory leaves the door wide open for a variety of paranormal phenomena to exist in a reality where our own matter operates in synchronized fashion with everything else.

In his book, *The Synchronized Universe*, Swanson points out that parallel universes superimposed upon our own can differ in phase or frequency of their synchronization. Basically, people can only access their own universal "sheet" of existence, and thus believe that is the only one that exists. The same goes for anything moving around on all the other parallel "sheet" universes.

But alter the phase or frequency just so, and an object can disappear from one reality, and appear in another. Similar to a ghost.

The synchronized universe theory could explain how UFOs can appear and vanish instantly, as reported in hundreds of sightings, and how teleportation might be achieved. It could also explain the existence of "subtle energy," which, Swanson theorizes, arises from the motions coupled across the layers of parallel universes. Subtle energies, Swanson states, are the "coherent structure which crosses several of these parallel realities, and therefore is 'higher dimensional.'" The interesting thing about synchronization is that it allows every particle a fundamental frequency proportional to its mass, and also explains how particles can become synchronized, similar to those Olympic swimmers, at small scales. This "synchronizing" of matter and energy would indeed allow for paranormal events to occur, and would no doubt explain why they might occur in transient, unpredictable ways. Ghosts, out-of-body experiences, remote viewing, psychokinesis, and poltergeist activity all may be the result of the synching of particles and matter between various levels of existence, creating a literal means for moving between dimensions of reality. Swanson states, "Paranormal effects and 'subtle energy' cause a synchronization across adjacent parallel universes. When this occurs, these adjacent universes become to a degree synchronized with ours. The interaction becomes more coherent, more in phase."

When there is no synchronization, we experience the other universe or dimension as "random noise."

Random Noise

Coherence might also reduce the random noise between systems, allowing more cooperation between two parallel systems, or universes. Swanson calls this coherence between parallel systems a "hyperdimensional" structure that crosses the dimensions. He suggests these may also be possible models for human consciousness and the concept of a soul.

A healer or someone who can move objects at will (or, in the case of poltergeists, not at will) can use the concept of synchronization, even if they don't understand what they are doing. Perhaps this explains the dynamic results of John Hutchinson's experiments in his lab, where objects moved about and levitated when exposed to a variety of electromagnetic stimulations. If just the right frequency was achieved, stuff went crazy. If not, the lab was quiet but for the hum of the machines.

Even teleportation fits into the synchronized universe theory. Swanson points out, "The behavior and position of matter is dependent on its radiation field, which keeps it in place, gives it inertia, and allows it to interact with other matter in the universe. If we shift the phases of radiation coming into the particle and coming out of it from the past and the future, we may be able to shift it's position." This may be the key to a form of "hyperdrive" that causes teleportation.

Physicist F. David Peat suggests that synchronicities are "flaws" in the fabric of reality, momentary fissures that offer those sensitive to them, a brief peek into the implicate, underlying order of nature. We know from the Law of the Conservation of Energy that the total inflow of energy into a system must equal the total outflow of energy from the system, plus the change in the energy contained within the system. In other words, energy never dies. It is converted into another form. Ghosts, if they represent trapped energy, may move between parallel dimensions or universes by this process of synchronization (possibly using the ZPF as a vehicle for moving between the dimensions) and become visible in our world because they are still energy. Many ghost sightings involve balls of plasma and changes in electromagnetic field measurements, as well as visible signs of energy manipulations. Lights flicker or pop, static appears on radios and televisions, and people report the feeling of hair rising on their skin or the back of their necks.

These are also widely reported elements of UFO sightings (stalled car engines, blackouts, phone interference, and so on). Obviously, energy is present, and affecting its environment.

In a paper titled "Searching for the Universal Matrix in Metaphysics" by Hal Puthoff published in *Research News and Opportunities in Science and Technology,* he makes a similar assumption in regards to the presence of Zero Point Energy as universal background energy. "Rather, to the degree that 'energy' is involved not only in the physical but in nominally non- or para-physical phenomena (including perhaps such 'mundane' phenomena as thought, charisma, etc., let alone psychokinesis), then such energy patterns might in principle emerge as a result of cohering or patterning the otherwise random, ambient zero-point energy."

Resonance

Again, one is reminded of music. When two or more notes are played together, they either create beautiful harmony, or noisy discord. The notes that work well together create a resonance. Those that don't "make sweet music" together create a dissonance. The sonic vibration or waveform, because sound is made up of waves, will either be in synch, or way off, and your ear, and body, can tell the difference.

Different types of music appeal to different people. I might love Rammstein, and you might love Madonna. We both might dig Mozart. Western music has a whole different "vibe" than the music of the Far East. Each inspires a different emotion and feeling, which is why many meditators, Shamans, and healers rely on music to help them transcend the normal state of everyday consciousness and enter other realms of reality. They literally "tune into" frequencies of reality that are beyond the range of our normal perception, the same way a dog hears sounds we cannot hear, and a bumblebee can perceive different light waves than us.

If paranormal events can only occur when the "right two notes are struck," then something in the brain, or the body, is responding to those notes and "tuning in" to the phenomena that is otherwise beyond perception. Or perhaps it is always there, but we rarely are in a state of mind to perceive it. This would explain why die-hard skeptics never have a paranormal experience. They just plain aren't "in tune."

Sound waves act like any other waves and vibrate to varying frequencies. Even magnetic waves vibrate, and most paranormal phenomena are associated with physical effects that imply the presence of energy, often electromagnetic, but sometimes even sonic energy.

Music serves as a great example for this theory of resonance, because it operates much like the universe itself does. To take it a step further, healers suggest that the human body is also a harmonizing machine, one that either is in synch with health or with disease. In his intriguing book *The Secret Life of Water*, Dr. Masaru Emoto photographed water crystals exposed to various emotions, thoughts, and intentions, and found stunning results. The photographs show remarkable differences in the beauty and symmetry of water crystals that are exposed to positive information as opposed to negative.

Emoto believes that resonance has a lot to do with how water is able to respond to someone's prayers, or even a nice piece of music. It's all about vibration. He even experimented with a man named Alan Roubik, an American pianist. Roubik bases his career on the belief that music can heal. Emoto had Roubik play a beautiful musical score, which was recorded and exposed to water crystals. Using a Magnetic Resonance Analyzer, which detects the minutest vibrations, they were able to take measurements of the amazingly delicate crystals that formed, showing how the water itself was affected by the music.

Emoto espouses "hado medicine." Hado is the subtle energy that exists in all things, such as the orgone energy of Wilhelm Reich, or "chi" energy. Emoto believes that the human body, which is made up of 70 percent water, responds as the water crystals do to music, prayers, and words, all of which are sound waves. In hado medicine, the body is likened to a universe of its own, with trillions of cells doing their specific jobs while "simultaneously harmonizing" with other cells. "The organs, nerves, and cells of the body have their own unique frequency," Emoto writes, and the body is like a grand orchestra that harmonizes the various sounds. When there is discord, or "*dis-chord*," there is disease.

Orgone Energy

Wilhelm Reich's "orgone energy" also operated on the principle of life energy that could either heal or harm the human body. His energy was considered the creative force in nature—present everywhere, constantly in motion, and the medium for electromagnetic force and even gravitation. Orgone, as Reich claimed, had no mass itself, and therefore could not be measured, but its effect could be detected in the movement of light waves or the force of gravity.

One of the characteristics of orgone is its ability to attract other streams of orgone and superimpose, which Reich claimed was the fundamental form of the creative process in the universe. Reich even devised machines to control and manipulate orgone, the best-known being the "orgone energy accumulator," made up of layers of metal and non-metal that served to concentrate the energy in the machine's interior.

Reich also created a weather control device, using orgone energy. The theory behind this device was that orgone energy could be extracted from the atmosphere using a directional antenna, then the energy could be redistributed to trigger changes in the weather.

Orgone could also be used for healing diseases, according to studies done with the subtle energy by Bernard Grad. In the early 1950s, Grad met Reich and felt that the concept of orgone, which was being ridiculed by other scientists at the time, held some promise that deserved further study. He used an orgone accumulator to test its effects on mice with leukemia. He used two groups of mice: the first had tumors implanted under the skin, the second had tumors develop spontaneously. Those mice that had been implanted with tumors showed little effect after use of the accumulator, but the second group showed a significant effect, with a 20 percent decrease in the incidence of leukemia. Other studies followed, but the most interesting find came when Grad met a healer named Estebany who compared the life energy he used to heal to orgone. Grad used the healer in experiments involving dying mice, resulting in a higher survival rate not only for Estebany's mice, but also for those given to another lab tech who also proved to have some healing ability.

These experiments by no means prove that orgone exists, or that it can heal cancer, but they, and other ongoing experiments, do hint at the presence of a field of energy underlying life itself, a field that can be manipulated and worked with for the promotion of physical well-being, as well as psi ability. It also parallels the use of chi or ki energy in the Eastern martial arts and wisdom teachings, where the balance and harmony of the "life energy" is essential to mental, spiritual, and physical well-being.

The Holographic Model and the Brain

Even our brains seem to have access to this field of all possibility. Neurosurgeon Karl Pribram spent decades pondering the secrets of the human brain and memory, and eventually discovered that the brain works somewhat similar to a hologram, and that when you first see something, specific frequencies resonate in the brain's neurons, which then send information about the frequencies to other sets of neurons, and the process continues until your neurons construct an image of what you are looking at. That the brain seems to process information in wave-frequency patterns suggested that human memory could hold amazing amounts of information "in storage," and, using the same holographic model, will be able to access and recall a memory as a three-dimensional image.

Pribram also proved to many skeptical colleagues that memory is distributed throughout the brain, rather than located in one particular system. He pointed to research using rats in a lab. The rats had major nerve pathways in their brains cut and some of their primary visual cortex removed, yet they could still remember and perform complex skills. The brain, similar to a hologram, stored every element of the original image over the entire system.

The holographic model of the brain could allow for ESP and telepathy, by the same process, according to Claude Swanson. "If the sender can cause energy or information to refocus at some other point in space/time using the 4-D holographic principle, then it can be received if a person is there to sense the thought-form." ESP experiments in lab settings have shown that, indeed, people get the best results when they are calm, and in a certain state of mind, as well as when the sender and receiver are "in synch," Swanson adds.

Pribram's brain research didn't stop with the holographic model and its implications for a variety of interesting abilities. He also showed that the brain acted similar to a frequency analyzer that literally filters out unlimited wave information from the Zero Point Field, where all the possible information existed. This allows the mind to take and use only what it needs, and not be overwhelmed with a bombardment of unnecessary frequencies trying to compete for some "brain time."

In *The Field*, Lynne McTaggart states that the human body's ability to exchange information with a mutable field of quantum fluctuations has profound implications. "The idea of a system of exchanged and patterned energy and its memory and recall in the Zero Point Field hinted at all manner of possibility for human beings and their relation to the world."

I Feel Like I've Been Here Before . . .

Think about the simple experience of déjà vu, which I am convinced is our ability to access, albeit in very small doses, glimpses of ourselves in a parallel universe or dimension, or even the ZPF. When you have déjà vu, you are remembering what is happening to you in the present moment. That doesn't make sense. If it is happening in the present moment, how can you have a memory of it? But if there really are parallel universes that have "branched off" each time we make a different choice, then it makes sense that we would, in many of those universes, be doing the same exact things at the same exact time.

When déjà vu occurs, most people immediately stop the flow by instantly stating to anyone standing next to him or her, "I am having déjà vu." Next time, try going with the experience, letting it just happen without resistance. I have done this and have had rather extended déjà vu experiences where I actually felt myself going out of my body until something outside of myself cut the feeling short (usually a whining kid or barking dog). If I exist in any number of other universes, I would no doubt find myself, in more than one of them at a time, tying my shoe while whistling "Jungle Boogie," or pretending to audition for *American Idol* in front of my bathroom mirror. (Not that I really do any of those things.)

What déjà vu is, then, is my mind's few successful attempts to find and synch up with another reality, even just for a minute or two. Or perhaps my mind is accessing the Zero Point Field, which, similar to Edgar Cayce's Akashic Records, supposedly contains the memories of every thought, action, or event that ever happened or will ever happen in the universe. Not all déjà vu, after all, is just a memory of something that already happened. Sometimes, you are able to predict what is going to happen next, suggesting that the information you are accessing is beyond the confines of our linear time.

Several large university research studies have tried to pinpoint the mechanism at work behind déjà vu. One such study, done at the University of Leeds in England, led to an outpouring of response from people claiming they have "chronic déjà vu," a condition that seems to be based upon the misfiring of information in the brain's memory system. The researchers at the Leeds study actually succeeded in finding ways to test déjà vu in patients in a lab setting, and are now launching a new project with the University of York in England to do neuro-imaging of the parts of the brain activated during an episode.

But the focus of these research studies is on déjà vu as a problem or a disease, rather than as a means of communicating across a spectrum of realities not always available to the human mind. In addition, the researchers never mentioned anything about precognitive déjà vu, instead focusing on past memory déjà vu, which could easily be passed off as the brain itself creating a false sense of memory where none existed before.

If déjà vu, dreams, and out-of-body experiences, including near-death experiences, allow us to wander for a while in other realities, it begs the question: Why can't we easily access these other realities all the time, at any time we please? I might suggest that it is because if we could, we would go insane from information overload. Our brains operate as information sorting machines, only bringing into our perceptive fields what we need to see, hear, and know to survive, and hopefully thrive. If we were able to constantly access other dimensions, universes, and levels of reality, we would not be able to function well enough in any one of them to survive. If it were so easy to move between worlds, our very survival would be threatened. We might never settle down, have a family, hold down a job, and pay our mortgage. It's hard enough to be homeless, unemployed, and living with your parents in one universe—imagine doing it in 10! Obviously, we humans were conditioned to live one life at a time. (Those leading double lives are exempt.)

Dreaming on a Different Frequency

That does not prevent us from being able to (when we meditate or go into a trance or zone out at our computer) have a peek into other dimensions of existence.

People do this all the time when they dream, when they know who is on the phone after only one ring, when they get a "bad vibe" about a person they've just met, or when they hear the voice of their dead grandmother telling them not to marry the jerk. The ancient art of dowsing, or using a stick or rod to find water sources beneath the ground, is said to involve the ability of the human dowser to sense, via the stick or rod, the presence of certain kinds of wave or particle emissions associated with the water substrate. Remote viewers may be tuning into a frequency upon which the information they are either sending or receiving is naturally "broadcasting," as a psychic might pick up on similar frequencies when doing a reading or having their own clairvoyant episode.

Electromagnetic radiation is but one of the possibilities that could serve as the playing field for paranormal experiences to occur upon. A telepathic impression or a past-life recall might just be the result of picking up on an already existing source of information embedded in a field of emissions.

If paranormal elements operate on a different frequency, again we must think of the dog that hears a high-pitched whistle that is soundless to us. Many species of snakes have organs that allow them to perceive the infrared range of the electromagnetic spectrum and "see" the heat of another living thing. Sharks and eels are ultra-sensitive to changes in electrical fields, and some insects can see ultraviolet light. These creatures are obviously observing things we humans simply aren't built to observe.

Yet somewhere there are humans who can hear these whistles and sense EMF alterations, just as there are humans who can see auras and other forms of light energy the rest of us cannot. It could just be a matter, then, of perception.

Or it could be that some humans simply know what frequencies to tune into.

Joshua P. Warren says in *How to Hunt Ghosts*, that when we look at a car, all we are really seeing is light reflected from the matter. The car appears to be solid, but is not. "It is a misconception that something composed entirely of light must be translucent. Everything we see is composed entirely of light, or the lack thereof. Conceptually, seeing a ghost is no different from seeing a living person." He compares the ability of a ghost, which could be "the residual, conscious energy left behind by a physically dead human," to operate much in the same way that magnetic lines of force do, able to interact with the physical world. "If we surround a magnet with iron filings, its field becomes visible three-dimensionally. The invisible form quickly emerges, almost like a materialized phantom."

Warren suggests that, just as the human body is made up of cells, the ghostly body is made up of free-floating static electrical charges called "ions." Ions are not visible, but you can still see and feel their effects on the environment. In the case of ghosts, that would mean hair raising on the backs of arms, cold chills, three-dimensional forms, and moving objects, even if you can never see the ghost, or the energy behind them, itself. "These discoveries . . . demonstrate that aspects of life can certainly exist in realms non-physical to ours. What we measure physically is only one layer of a world with nearly infinite levels." In other words, Aunt Mildred may indeed be dead in one universe, but alive and kicking and able to pop over for a spot of tea in another.

Whether energy is synching up, emerging in a coherent pattern from the ZPF, moving along a morphogenetic field, or even floating in a sea of quintessence, there is definitely fertile, dynamic ground for the movement of energy between our world and the world of the paranormal (which could in fact be our own world, or one parallel to ours). If indeed there are ways matter can disappear from one state of existence, only to reappear in another, then it begs one very important question: Why don't we see more ghosts, UFOs, and psychic activity on a larger scale? Why are we not hearing wives all over the world shouting at their bewildered spouses, "Honey, did you leave the door to the 10th dimension open? Here comes your dead Uncle Arthur *again!"*

We are limited to our five senses: sight, smell, sound, touch, and taste. But what about this sixth sense? Perhaps that is the sense, underdeveloped in most, yet present in all, that allows us to turn on and tune in to the frequencies that reveal the other layers of existence. I think of years ago, long before cell phones, when I owned a CB radio and loved to listen to motorists and truckers chatter. Occasionally, there would be some bandwidth cross talk from a nearby Ham radio operator. Ghosts, UFOs, and other paranormal events may be cross-talk cutting into our normal bandwidth—chatter from another dimension, universe, or somewhere in between. We don't normally hear it because it is usually operating on another frequency altogether, but sometimes signals get crossed.

What is the mechanism that links these various fields, forces, synchronized seas of resonance, vibration, matched scales, pitches, and frequencies to our visible experience of reality? What is the door between the implicate and the explicate order of things? What is the one thing that stands between the normal and the paranormal, the natural and the supernatural, the science and the *psi*ence?

I'll give you a hint. If you are alive and breathing and reading this book, you have one.

Cosmic Mind, Conscious Mind:
Consciousness, Perception, and the Power of the Observer

Consciousness...is the phenomenon whereby the universe's very existence is made known.
> —Roger Penrose, *The Emperor's New Mind*

The universe begins to look more like a great thought than a great machine.
> —James Jeans, *The Mysterious Universe*

You can observe a lot just by watching.
> —Yogi Berra

The next time someone tells you "it's all in your mind," you may want to compliment him or her on his or her scientific and intellectual acumen. Once the domain of religion, philosophy, and metaphysics, the power of consciousness and perception is becoming more obvious to scientists plumbing the depths of quantum physics. It seems that even on the level of the very, very small, the role of the observer is tantamount to the outcome of the results of an experiment.

It might, in fact, be the very foundation of reality itself.

Recall the famous words of French philosopher, mathematician, and scientist René Descartes. "I think, therefore I am." Descartes recognized that the act of thinking suggests proof of existence, of self-awareness. But I suggest that, for the sake of this book, we take that sentiment one step further to explain the role of the observing, conscious mind in the macrocosmic and microcosmic.

"I observe, therefore you exist."

We know from experiments with photons and other particles on the quantum level that the observer changes the outcome of the results. Think of Schrödinger's cat-in-the-box, or the paradox of the dual particle-wave nature before the wave function collapses. Until the moment of observation of the particle, there is no real particle to observe. Physicist Paul Davies writes in his book *Superforce* that "In the absence of an observation a quantum system will evolve in a certain way. When an observation is made, an entirely different type of change occurs. Just what produces this different behavior is not clear, but at least some physicists insist that it is explicitly caused by the mind itself." This statement was written in 1984, and since then, more physicists and scientists have conceded the role of the mind in creating the physical reality we "observe" on a daily basis. More and more research is coming to the forefront every day that confirms that the outcomes of quantum phenomena can be modified by consciousness.

Consciousness of Physics

What is consciousness? A quick look at the dictionary turns up a variety of answers. "Awakened state; the awareness of self and of what one is doing and why; the totality of one's thoughts, feelings and experiences; a state of knowing the self; cognizance." Consciousness, then, is a state of awareness, of knowing that one exists. We possess consciousness when we are awake and aware of our own existence.

Consciousness is also described as the ability to think and perceive. Thoughts, perception, and consciousness are the key players in understanding the connections between the paranormal and the normal, just as they are the key players in determining the outcome of a quantum physics experiment. The nonlocality experiments of Alain Aspect and his team in 1982, discussed in Part II of this book, proved that space is "non-local" and that the world is not made up of separate objects that, when put together, make the universe as we know it. Instead, these ground-breaking experiments showed that the "observer" and the "object being observed" were connected and part of an indivisible whole, with everyone and everything affecting and influencing everything else.

Some physicists look to the realm of "superspace," or the Zero Point Field as the home of the Cosmic Consciousness that observed, and continues to observe, the universe into being. Linked to this cosmic mind is the mind of the individual human, who, on a microcosmic scale, observes his or her own reality into being. And as the previous chapter shows, synching of vibrations and frequencies, whether those in the electromagnetic field or the human brain, can certainly lead to an experience of something beyond our normal, waking consciousness such as a precognitive dream, or a ghostly encounter.

Okay, I can hear you all throwing down your books and running away, shouting something about "religious mumbo-jumbo" and "metaphysical nonsense." But before you burn my book and label me a heretic (it wouldn't be the first time!) bear with me a bit longer.

Physicist Amit Goswami, author of *Self-Aware Universe: How Consciousness Creates the Material World*, believes that the material world of quantum physics is only possibility, and that through the conversion of possibility into actuality, "consciousness creates the manifest world." He suggests the universe is self-aware, but it is self-aware through us, through our individual consciousness. Goswami is one of many physicists working in the cutting-edge field of "New Science" or the science of consciousness and its role in physical, manifest reality. He points to the Heisenberg Uncertainty Principle as the beginning of the "consciousness revolution" that based itself up on the idea that the observer affected the observed and launched a new direction of research that combined the world of matter with the world of mind.

Information theory also accounts for the role of consciousness. In fact, theorists suggest that the process of quantum superposition and collapse is similar to the way the mind works. Charles Seife, in his book *Decoding the Universe* states that ideas seem to start out in the superposition in the preconscious, then "wind up in the conscious mind as the superposition ends and the wave function collapses." He points to research done by British mathematician and quantum theorist Roger Penrose, who wrote about the similarities between the human brain and a quantum computer in his book, *The Emperor's New Mind*. Penrose and colleague Stuart Hameroff, an anesthesiologist, worked with "tubulins," tiny tubes constructed out of protein that may be the seat of the brain's "quantum nature." These tubulins act akin to particles in a state of superposition, able to exist in both an expanded and contracted state, which might also affect the state of neighboring tubulins, much as a quantum computer. Penrose and Hameroff also proposed that consciousness might even reside in this "quantum brain."

Other scientists, such as Max Tegmark at the University of Pennsylvania, suggest the brain is not a good environment for quantum computing, pointing to the rapid decoherence of the superposition and the inability of neurons to respond quickly enough to the information. Seife reminds us that even if the brain is one big information-storing machine, "it is so complex and intricate that scientists have no real idea about how it does what it does except in a very gross manner."

Interestingly, many theorists propose that the act of decoherence is what prevents two different branches of the multiverse from interfering with one another. But as physicist Brian Josephson explains in his paper "The Paranormal: The Evidence and Its Implications for Consciousness," written with colleague, professor of statistics Jessica Utts, decoherence still "does not answer the question of why our experience is of one particular branch and not any other. Perhaps, despite the unpopularity of the idea, *the experiencers of the reality are also the selectors*. This concept, Josephson and Utts add, suggests that a deeper "subquantum domain" exists, whereby a psychic can "direct random energy at the subquantum level for her own purposes." Again, the role of the observer, the selector, and the experiencer, plays a large part in the manifestation of reality, including the presence of psychic, or psi, phenomena. Indeed, this would explain why some people seem to "have it" and some don't when it comes to psychic ability.

Consciousness is the key, and the observer chooses the branch of experience. Josephson also suggested in another paper, "String Theory, Universal Mind and the Paranormal," that "some aspects of mentality involve a realm of reality largely, but not completely, disconnected from the phenomena manifested in conventional physics." He points to the informational aspect of life; there is a biological character involved with the informational processing of organisms, one that sees life "able to shape its environment in a partnership with it." Two life forms sharing their mental states could account for such phenomena as ESP and telepathy.

Josephson describes a potential "shared mental bubble," the contents of which are available to both life forms involved. "The point to bear in mind," he continues, "is that in the biological realm the phenomena that manifest are governed not only by what is physically possible, but also by which of those physically permitted possibilities are likely to be of overall benefit to the organism concerned."

When I read that quote, I thought of the work done by many serious parapsychologists who suggest greater levels of psi phenomena are reported under laboratory testing conditions when the research subjects have a *vested interest in the outcome of the experiments.* Implied meaning can often make a big difference. Richard S. Broughton, scientist and former president of the Parapsychological Association, wrote in his presidential address in 1987 about the importance of meaning when working with test subjects. He proposed that researchers ask the question "whom does psi serve?" when working with subjects, pointing to the "need-serving" nature of psi phenomena and how it might affect the skill levels of individuals.

Obviously, if there is a greater need, often survival-based, for psi, then the psi will show up in a greater amount in the subject. Broughton stated, "Quite a few experiments in ESP and PK can be read as providing support for both the need-serving character and its operation at an unconscious level." Test results seem to imply that there are psychological and biological implications to paranormal phenomenon, and that their functioning significance must be taken into account.

Thus, if one person has a greater need to display PK, they will display it. But that does not mean psi only shows up when there is a desperate need for it, Broughton continues, but that researchers must keep in mind that "probably the primary function of psi is to help the individual survive when faced with serious threats to health and safety, and to gain a competitive advantage in the struggle for survival."

Do we, then, have a choice to be psychic or to see a ghost? Perhaps we do, but on a subconscious level. And when that choice is made on a conscious level, can our brains then "synch" with the exact frequency necessary to perceive what is considered beyond normal perception? PEAR, the Princeton Engineering Anomalies Research laboratory, is deeply involved in studying the idea that thoughts can affect the exterior world, and that consciousness can influence other objects and organisms. Its work into remote viewing, and the power of conscious thought to affect random number generators, is available for viewing on its Web site, *www.princeton.edu/~pear/* and strongly points to the ability of some subjects to transmit and receive information via the mind alone. In a PEAR paper titled "A Modular Model of Mind/Matter Manifestations," the researchers have created a model that proposes mind and matter unite with and influence each other in the deep levels of the unconscious. In other papers, they extend their research to conclude that we can alter our consciousness to therefore alter the filters of perception we use to "see" and experience our reality.

Russell Targ and Elizabeth Rauscher worked with researchers around the world to carry out remote viewing experiments that convinced them of the evidence for a "mode of perception, or direct knowing of distant events and objects." Their findings, summarized in *The Speed of Thought: Investigation of a Complex Space/time Metric to Describe Psychic Phenomena* and available on Targ's Web site, propose a geometrical model of space/time that is eight-dimensional and allows for a connection of zero distance between points in the "complex manifold." This model, they state, describes the major elements of experimental parapsychology, yet is consistent with the structure of modern physics.

Rauscher and Targ delved into the connections between our awareness and our use of psi abilities, focusing on data gained from over a century of laboratory experiments into various phenomena, starting with the usual objection of science towards psi—the fact that it appears to conflict with the known laws of physics. Their theoretical model finds areas of agreement between physics and psi, using six spatial dimensions and two temporal coordinates. They present their "metric of complex eight-space" as the measure of the manner in which one can "physically or psychically move along a world line of space and time." For remote viewing and precognition to work, they state, the experienced distance between subject and target can be zero, and their proposed eight-space can "always provide a path, or world line in space and time, which connects the viewer to a remote target, so that his awareness experiences zero spatial and/or temporal distance in the metric."

The authors believe psi abilities are "fundamental to our understanding of consciousness itself." In fact, they believe that psi functioning might even be the means by which our consciousness makes itself known to the internal and external world, and to our selves, our own individual awareness. They point to experiments involving hundreds of subjects in the areas of precognition and remote viewing, where results showed strong evidence for the existence of knowledge about the future. Some of the experiments were those of Dr. Dean Radin, a Senior Scientist at the Institute of Noetic Sciences.

Radin, author of *The Conscious Universe and Entangled Minds*, performed experiments measuring the "orienting response," or the "fight of flight reactions" of subjects exposed to pictures that were scary or distressing. The interesting thing is, the physiology of the person viewing the picture actually changed *a few seconds before they saw the picture*, showing a direct correlation between these physical reactions and premonition—the inner knowing that a future event is going to occur. Somehow, our bodies, or our minds, know when something is going to happen, implying that our ordinary perception of time is not complete. Intentions somehow have the ability to transcend time and literally work backward to influence the past, or forward to predict the future. Can our own future self be influencing us, helping to make our decisions that come in the form of premonitions?

Experiments with random number generators show that prerecorded random bits of information actually conformed to intentions produced in the future when a person is asked to try to influence them. The more people trying to influence the machines, the greater the effects. We recall the "black box generators" mentioned in earlier chapters, where machines sensed great world events before they happened. The Random Event Generators (REGs) may be the closest thing we've come to proof that predicting the future is possible, and that the influence of human consciousness is a mandatory part of those predictions. Think about the September 11, 2001, terrorist attacks in New York and Washington, DC, and how many people must have had precognitive dreams, visions, and just hints of something terribly amiss. Many stories emerged afterward of people who deliberately stayed home from work to avoid the Twin Towers, missed planes that later crashed because they felt a dire inner warning not to get on, or were angry about being late to work, only to later find they were blessed with another chance at life. The REGs recorded shifts in the patterns of numbers *four hours before* the first two planes hit that morning. How many human REGs recorded similar shifts in normal patterns? One has to wonder how many of the victims of those attacks had similar feelings and hunches and intuitive urges, but just did not make the "psychic connection," to quote a line from *Close Encounters of the Third Kind*.

Most of the time, we ignore these intuitions, gut feelings, even those outright voices shouting "danger, Will Robinson" in our heads.

Called "presentiment" according to Rauscher and Targ, this inner knowing of the future suggests that the future may already be determined, and that because our awareness is nonlocal, the past and the future may act as attractors pulling the present toward either. We don't lose our free will, though, because we can, as they suggest, "use our premonitory information to make even more informed decisions about what we should be doing." That would explain the many stories of people who had a feeling about getting on a plane, and choosing not to, only to later discover that the plane crashed or exploded mid-air.

The human brain acts as an amazing filter, accepting and rejecting information and stimuli based upon our survival needs. You could say we operate on a strictly "need-to-know" basis, with our brain acting as the sentry at the gate, refusing entry to anything that we just plain don't need to understand. People who are able to shift their perception thus shift their "need-to-know" basis to include *things they did not perceive before.* If you don't need to see it, you won't see it.

On the other hand, phenomena such as ghosts, UFOs, and other entities that take on a nearly-physical form may be coming into our consciousness without our consent, simply because they have found the

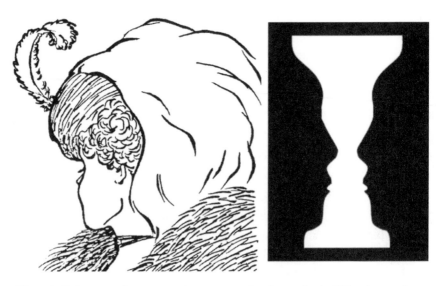

The mind chooses what to perceive on a need-to-know basis. What image does your mind have a "need-to-know"?

operating mechanism for being able to do so. And on some level, we must be consenting to perceive their presence, just as we tend to see, or not see, what we want to see in daily life. Think about the conditioning behavior we exhibit, where we are not consciously aware of why we repeat a bad habit or make the same mistake over and over again. Just as we are blind to parts of our own behavior and identity, we are blind to parts of the "spectrum" of reality.

I can't help but recall the quote by William Blake, "If the doors of perception were cleansed everything would appear to man as it is, infinite."

About.com's guide to Paranormal Phenomena Stephen Wagner told me that he feels the scientific method fails when applied to psi. "The scientific method demands repeatable results. This is not always possible. A person in a 'haunted house' can see an apparition or experience a telekinetic event one night, but when the researchers come in with their equipment, nothing happens."

I add that if a person's perception and level of consciousness are key factors, which quantum physics supports, then there is even less likelihood of being able to consistently repeat a desired outcome. Wagner told me about a ghostly experience he had at the age of 20, when he kept hearing the distinct sound of breathing when he went to bed at night. Despite repeated attempts to find a rational explanation, he found none (the breathing was not his and he was alone at the time). Eventually, it stopped, and never returned.

I have felt the distinct presence of my beloved dead cat, Elvis, at the foot of my bed. It happens on some nights, not on others. These fleeting experiences occur all over the world, every day to a wide variety of people, and may only happen once. I recall many people who have had psi experiences say something to the effect of "I guess the timing was just right," or "there was a certain feeling I've never had before, or since."

People who experience near-death experiences (NDEs) and out-of-body travel are also experiencing a fleeting glimpse into altered states of consciousness that may, or may not, be voluntary. NDEs are certainly not repeated over the course of the experiencer's lifetime, just the one time it is "needed" or holds meaning. Out-of-body experiences often cannot be controlled, although those who have mastered them probably do so by shifting their consciousness to accept the experience in the first place. Doctors working with patients who experience both NDEs and see ghosts of dead loved ones waiting to embrace them suggest that other levels of consciousness are being accessed. Dr. James L. Hallenbeck of the V.A. Palo Alto Health Care System compares it to radio frequencies in his recent book, *Palliative Care Perspectives*, stating that "in normal wakefulness, we function and interact on a relatively narrow and shared frequency that allows both transmission and reception of shared frequencies. When patients at the end of life experience altered states, it is as if their radio frequency, their wavelength, has shifted." And it's all a matter of just tweaking the dial ever so slightly.

Consciousness creates reality, according to many mystics, and some scientists are even jumping on the bandwagon, agreeing that our consciousness has a special ability to help shape the manifest world of our experience. Tweaking the radio dial can lead to greater understanding of the sheer extent of reality, and, if we have a collective consciousness, as many mystics believe, we could be manifesting experiences that are shared, or contagious, when more than one person tunes into the same alternate frequency. The illusion of separation of space and time leads us to believe we, too, are separate, but look at how germs and viruses are able to "infect" one body and jump to another and another, until an epidemic occurs.

Shared Experience

I refer to the theory of shared experience as "contagion." When an idea or intention is shared by enough people to create a critical mass, or as British author Malcolm Gladwell put it in his best-selling book, *Tipping Point*, we begin to see a wave of UFO sightings in a particular region, or a series of bizarre poltergeist events in a family unit or even a neighborhood. Contagion is most likely the cause behind mass demonic possession, hysteria, and other forms of collective belief. Even cults, religions, and politics operate on a level of contagion. When enough people believe something and accept it into their consciousness, they manifest it as a reality.

One chilling example might be the "leaping" demonic possession of the Ursuline nuns of Loudon, France from 1632–1634, which are among the most famous examples in the history of diabolic possession in early modern Europe. It began on the night of September 22, 1632, when two nuns encountered a spirit. Two days later, a large black ball crossed the refectory, pushing some nuns to the ground. The following week, a human skeleton was seen walking in the convent's corridors. In the following weeks, numerous nuns heard voices and were beaten and slapped by invisible entities. Exhibiting supernatural physical strength, screaming, crying, fainting, and suffering from uncontrollable seizures and convulsions, the sisters showed all the traditional marks of diabolic possession. According to well-established Catholic tradition, the local clergy then organized exorcism rituals.

It was during these ceremonies that the nuns accused a controversial local priest named Grandier of having signed a pact with Satan, a claim supported by the demons speaking through the sisters' mouths. Following a series of trials, he was found guilty and executed on August 18, 1634. The possessing demons then began to depart from the nuns' bodies, and the town of Loudon returned to normalcy once again.

Whether these nuns were actually providing temporary corporeal housing for demons, or just engaging in a contagion of mass hysteria, the end result was the same. Although some historians would later claim these women were reacting to the witch-hunt atmosphere of the times, and using these "possessions" to conveniently dispose of the nastier men in their lives, parapsychologists point to the numerous physical manifestations that often accompanied possession of activity similar to that associated with poltergeists and ghosts—moving objects, levitation, plasma, balls of light, and spectral beings that instantly appeared and disappeared and even walked through walls.

Contagion is present in something as common as an ordinary bad mood, which often spreads to those around us, just as a smile can "infect" others. Energy has the power to attract similar energies, and so, thoughts and emotions as well. Many Eastern philosophies use this concept at the very core of their healing practices, understanding that positive and negative thoughts, energies, and influences can affect the health of the body. If there is indeed a chi life force, our thoughts and beliefs in the form of our consciousness expressing itself inward and outward will definitely play a role in shaping our health. This goes for our emotional and mental health, as well as our physical health.

But contagion can do more than just manifest waves of feeling, belief, and emotion in groups of people, just as it can do so in groups of cells and organs within one human body. It can also affect how life "looks" to people in a physical, tangible sense.

Dr. Jacques Vallee believed that consciousness could create UFO sightings. In *Revelations*, he proposed that UFOs may not necessarily be visits from space travelers, but rather a manifestation of a more complex technology that acted similar to consciousness, manifesting outside the body from somewhere beyond space and time. He even believed the key to understanding UFOs lies in the psychic effects often produced during sightings. He also likened some UFO behavior to that of a holographic image, where the machine projecting the actual image remained outside the "view" of the conscious observer. We see only that which is being projected, as if we exist in a virtual universe made up of shadows, where the solidity of matter becomes more of an illusion.

Vallee took his speculation further, proposing that there was a "spiritual control system for human consciousness and that paranormal phenomena like UFOs are one of its manifestations." He urged scientists and researchers to focus their studies on the presence of a "system around us that transcends time as it transcends space." This system might be located in "outer space," but had the ability to manifest as physical objects that "could not be understood apart from their psychic and symbolic reality." He even went on to suggest that consciousness is "no longer simply a local function in the human brain," but rather the process by which informational associations are retrieved and received. Paranormal phenomena would then be expected, even common, as natural aspects of the "reality of human consciousness."

Paranormal researcher Nick Redfern shared his ideas with me about how we might be creating things such as UFOs, aliens, and various cryptozoological creatures out of our collective unconscious. This may serve a deep-seated need for human evolution, as did the myths of ancient times, and even the "faerie creatures" of the not-too-distant past. Our need to connect to the world on a deeper level may play a role in the manifestation of psi phenomena. Redfern points to the connection between UFOs and mythology of various cultures. Modern ufology is only about 60 years old and has changed drastically from the flying saucers and humanoids of the 1950s to the greys of the 1960s. Today we see sinister MIBs with a close connection to the paranoia and fear of terrorism that marks our current decade. Redfern then reminds us that in some small English towns and villages, the mythology and culture hasn't changed much since the 1950s, and thus manifests in the kinds of psi phenomena they report.

This idea that UFOs and mysterious creatures could represent "certain motifs that don't change over history, but just have different origins," as Redfern puts it, mirrors the research of Jung into archetypes and the collective consciousness and unconscious. We might be playing a role in bringing these phenomena into "existence," or, if they already exist, we are certainly providing the necessary doorway for them to come through. Our consciousness is that doorway.

Certainly, some UFOs are nothing more than our government's (or another country's government) attempts to hit the skies and soar, but there are way too many UFO cases involving a deeper level of reality, and a hint of interdimensional origin that go far beyond what any top secret military base may be fiddling around with. Redfern, in his many decades of research, found that in terms of the average "grey" alien, "it is also possible that we may actually have the power to influence how it (dimensional intelligence) appears . . . . People are conditioned to believe that this is what aliens look like, and so this is how these 'things' appear to us. Their appearance is determined by our expectation of what they look like." When we the people stop believing in a certain type of entity, such as little faeries and longhaired contactee-style aliens, then we the people stop seeing them.

Contagion works both ways. Our collective minds can infect others to see something . . . or not see it. Just as with the spread of disease from one physical body to another, paranormal experiences (and expectations!) may spread in a similar style from one consciousness to another. Denial is not a river in Egypt, but actually a strong form of contagion that we see evidence of every day in our news media, and in our collective bad habits and negative behavior patterns. Ever wonder why people keep doing dumb things even after they know better? If enough people still believe the dumb things they are doing are, well, *not dumb*, the collective "mind" continues to manifest the results, despite the efforts of the well-intended few that have awakened to a greater truth. Critical mass occurs without judgment, suggesting that the unified field that underlies all things is neutral, much to the chagrin of the deeply religious.

F. David Peat, in his book *Synchronicity*, talks about the objective intelligence or "creative ordering" existent in nature. "Nature contains archetypal patterns and symmetries that do not exist in any explicit material sense but are enfolded within the various dynamic movements of the material world. Matter, accordingly, to such a view, does not represent a 'fundamental reality' but rather is the manifestation of something that lies beyond the material domain."

Consciousness on All Levels

Maybe Bohm's implicate order is simply the consciousness of the body of explicate order. The thought behind the manifestation.

Jose Arguelles, an art historian, calls it "salsa de vida," the dance of life, between people, between minds, between consciousnesses.

Sir Roger Penrose once proposed that consciousness was what happened when the brain's neurons reached a threshold of superposition, and then collapsed, as particles collapse when observed. Divided into preconscious, unconscious, and subconscious levels, this lower arena is where information flows and is then processed or "perceived" by the conscious level. Just as Schrödinger's cat is dead *and* alive before you peek in the box, the various levels of the conscious mind are open to anything until the neurons collapse and the conscious "sees."

Dr. Stuart Hameroff, Penrose's colleague, believes that consciousness exists "on the edge between the quantum and classical worlds . . . in that there is a universal protoconscious mind which we access, and which can influence us." In the book *What the BLEEP Do We Know!?*, based upon the successful 2004 independent film of the same name, Hameroff talks about the fundamental level of the universe as "a vast storehouse of truth, ethical and aesthetic values, and precursors of conscious experience, ready to influence our every conscious perception and choice. We are connected to the universe, and entangled with everyone else through this omniscient omnipresence, a sea of feelings and subjectivity." Our ability to use free will, or intention, is directly related to our ability to create a certain reality for ourselves. Dr. Joe Dispenza, also interviewed in *What the BLEEP*, says "all those experiences shape, neurologically, the fabric of what's taking place in our perception and in our world." Hameroff takes our role as observer and intender one step further, saying that if we are mindful of our conscious choices, we can be "divinely guided." In a sense, we either consciously tune in to the deeper implicate order, or we tune out, and experience a sense of discord or chaos.

Inspiration—the act of being "in spirit," may be what happens when we tune in to that deeper order underlying reality itself, thus leading to a range of successful inventions, artistic achievements, technical revolutions, and evolutionary leaps in understanding.

This implies that our consciousness is not only responsible for shaping our perception, but that it is responsible for creating our reality. A tree may still exist whether we walk in the woods and see it or not. But then again, as experiments with the collapsing wave function of quantum particles shows us, it may not exist *as a tree* until someone comes along and collapses the wave function and observes it *as a tree*. The very act of observation turns a mush of potentiality into a redwood, an elm or a maple. Psi, as with that tree, may either be a total creation of our conscious connection to a deeper order, say the Zero Point Field. Or it may already be there, waiting for us, and we just need to learn how to perceive it. Physicist Peter Russell says in *From Science to God: The Mystery of Consciousness and the Meaning of Light* the following: "Ultimately, there are as many ways of perceiving the world as there are species of life in the universe. What we take to be reality is simply the particular way the human mind sees and interprets the physical world."

Okay, so it really is all in your mind. The mechanics of neurons in the brain work in such a way that, literally, we can only perceive that which we can conceive of. When we perceive something, electrical impulses in the brain are carried to dendrites, which act as receptors. To register an object or experience as reality, we must have neurons firing and creating connections from dendrite to dendrite across microscopic gaps called synapses. The more complex the experience, or unique the object, the larger the dendrites and greater the synapses. This expanded synaptic activity allows our brains to process things it has not seen, or experienced, before.

For the many skeptics who have never seen a ghost or had a psychic experience, their brains have been literally hard-wired not to perceive what they cannot conceive of. A person who has never seen a ghost, and then "perceives" one, whether voluntarily or not, has literally stretched their brain to accommodate the new experience.

The same could be said for spiritual and religious experiences, which expand the neuronal activity in the brain, and have even been duplicated in a laboratory setting where certain parts of the brain have been stimulated to create a sense of "rapture" or an ecstatic state. Clearly, there is a connection between the brain itself and paranormal experiences. Consciousness itself may be the required key that opens the door to the brain's ability to perceive something it never has before. Similar to a hologram, the brain, via consciousness, sees the projection of reality onto a three-dimensional surface, and then creates meaning from that projection.

Many experiments have been conducted that show the extent to which the information obtained by the observer is indeed dependent on the observer's own assumptions and preconceptions, as well as their values. Reality has no meaning until we attach one to it, and that very meaning may become our reality, as in a spiritual or religious experience, or a UFO sighting that opens our minds to a new way of thinking about our world (one that we simply were not open to before).

The idea of a meaningful, spiritual conscious connection to the universal "mind" is not new to science or quantum physics by any means. In fact, those who believe there is nothing spiritual at all about the mysterious hidden workings of the multiverse might be surprised to find out that ancient wisdom has had a big jump on modern science for a long, long time.

As Above, So Below:
Ancient Wisdom and Modern Science

All theory, dear friend, is grey, but the golden tree of actual life springs ever green.
— Johann Wolfgang von Goethe

I would rather live in a world where my life is surrounded by mystery than live in a world so small that my mind could comprehend it.
— Harry Emerson Fosdick, Riverside Sermons

The fairest thing we can experience is the mysterious. It is the fundamental emotion which stands at the cradle of true art and true science. He who knows it not and can no longer wonder, no longer feel amazement, is as good as dead, a snuffed-out candle.
— Albert Einstein

If science is about the structure of the universe, then spirituality is about the essence. We can never really understand the nature of reality unless we find a theory that encompasses both the implicate and the explicate. The macrocosm and the microcosm. As above, so below.

Dorothy couldn't understand Oz until she saw what, or who, was behind the curtain pulling the strings.

Whether or not you believe in anything spiritual or religious, you can appreciate the depth of wisdom in ancient religious traditions whose writers and creators pondered the same questions about the universe as we do today. From the dawn of recorded history, humans have whispered of what lies behind the curtain, beneath the shrouded veil of everyday reality, beyond the door of ordinary perception and experience. They wondered about how it all worked, what laws were present and what fundamental truths governed the movement of stars and planets and even the atoms that made up all matter.

The Beginning

Heinz Pagels, in his book *The Cosmic Code: Quantum Physics as the Language of Nature,* said "I think the universe is a message written in code, a cosmic code, and the scientist's job is to decipher that code. This idea, that the universe is a message, is very old. It goes back to Greece . . ."

This coded message is behind both the scientific quest for a Theory of Everything, and a spiritual quest for the understanding of truth. From the ancient Egyptians to the Babylonians, Sumerians and Maoris, the creation myths speak of how matter arose from a field of energy, vibrating like a sea of light, giving birth to all manner of forms. The opening line of Genesis in the Hebrew Old Testament states that creation was brought forth from a formless void, and that the first order of business was the existence of light above all else.

We know from previous chapters that physicists believe in a field that permeates all of space, whether it is the Zero Point Field, or the ether, and that this field literally vibrates with energy. Light is the basic building block, so to speak, of all we see in the universe. *The Shvetashvatara Upanishad,* one of the oldest primary texts of the Hindu Vedas' mystical teachings, states, "The One 'I am' at the heart of all creation, Thou art the light of life." Light has, in many ancient and modern religious traditions, been associated with the prime life force, which is why we call spiritual masters "enlightened" or "illumined."

Peter Russell writes in *From Science to God* that light is fundamental to the universe, and "the light of consciousness is likewise fundamental. Without it there would be no experience." He goes on to ask if physical reality and the reality of the mind might share the same common ground—light.

The association of light, consciousness, and creative force, or God force, is present in most wisdom teachings. The field of energy from which creation sprang forth is often portrayed as the "mind of God" or a singular, "first mind" or "first thought" that manifested itself into matter and form through its own consciousness. From this primal consciousness all life arose, and to this primal consciousness, all life will one day return.

In Egyptian creation myths, this primal field was thought to consist of the "waters of life." The story of Atum, the high god of Egyptian mythology who, along with his descendants, created the universe, tells of a primordial ocean called "Nun," from which all basic matter was created. Nun was infinite and filled every dimension of space/time, and there was nothing else but Nun, also called the "father of the gods."

Physicist Paul LaViolette compares the primeval waters of creation to the "transmuting ether" of his theory of subquantum kinetics. "The primeval waters are invisible and of infinite extent, and their substance forms the basis for all the creative Universe." This ether-like field exists initially in an inert state, until Atum, the "Becoming One" emerges and the Nun becomes transmutative, giving birth to visible creation.

Throughout the ancient creation stories and holy texts, we find references to the same fundamental forces quantum physics and theoretical physics propose as the known laws of the universe. These laws were known to those in ancient times in the behavior of their gods and goddesses, in their myths and stories, and in the basic understanding of science they did possess. One only has to read the story of the god Pan-Ku of the ancient Chinese to see a correlation with the Big Bang. Pan-Ku, often depicted in Chinese art as a dwarf-like man, started out as an infinitesimal being no bigger than a speck of dust. He proceeded to grow in size for 18,000 years, carving from his body the universe and all its manifestations. His body became the Earth, and his blood, the oceans. His breath was the atmosphere itself.

The Japanese offer the *Kojiki*, a lush account of how everything began. Twin deities, Izanagi—the male who invites, and Izanami—the female who invites, appeared on a Floating Bridge of Heaven. Under this bridge was a field of mist, from which they created an island they could call home. On this island, they copulated and brought forth, after a few faulty trials, the eight islands of Japan, followed by the oceans, the Earth, and its winds and seasons. They also gave birth to a child, a spirit of Fire who consumed his mother at birth. In the throes of her death, she vomited a number of beings, and then died. Her mate wept for her, and from his tears sprang forth more of creation, and so on and so on.

Numerous other creation myths talk of a mysterious ocean or mist or sea of energy from which matter is made manifest. To many of the ancient peoples, this sea was alive, and that the act of creation was initiated by a sound, a word, or a "commandment of the tongue." This corresponds to the Hebrew Bible's "In the beginning was the Word," the "Logos" of creation, and suggests that sound itself had the power to form matter, even preceding light in order of existence.

Light, sound, field of energy, first mind. Consciousness. Ancient Sanskrit texts refer to *Purusha*, or Supreme Consciousness, and *Chittam*, mindstuff. These two things form the nature of all reality, the Chittam springing forth from Purusha in various grades and orders, from the minerals and plants to the animal kingdom and, finally, humans.

Paramahansa Yogananda, author of *Autobiography of a Yogi* and a great spiritual master to many followers, spoke about his experiences with yoga as coming into a "oneness" with that Supreme Consciousness, or God. He described being opened up in such a way that he felt as though the entire cosmos existed within his being. Perhaps he was experiencing a sense of what quantum physicists call "nonlocality"? Or maybe he had tapped fully into the Zero Point Field, where his consciousness merged with that of every other living thing across dimensions of space and time. Interestingly, Yogananda also describes the underlying field of energy from which matter is created as being composed of light. That energy, similar to the Zero Point Energy that could one day power spaceships to other universes, has gone by many names—*chi*, or *qi* of Taoism, the *prana* of yogic traditions, the *pneuma* of Greek mysticism, the *ruah* of the Hebrews, the *barakah* of Sufism, the *spiritus* of Latin, or the *Holy Spirit* of Christianity.

Accessing Consciousness

Yoga and meditation have long been suggested as the best ways of accessing higher levels of consciousness. Eastern traditions recognized the need for quieting the "outer" or explicate mind in order to access the "inner" or implicate mind, where that connection to the first mind, or Divine Mind, was to be found. Prayer and contemplation also serve to quiet the mind, and many adepts used these techniques to not only experience a higher level of conscious awareness, but to also access "paranormal" abilities such as precognition. Think of the Oracles at Delphi, and the prophets of the Old and New Testament, many of whom experienced trance-like visions of future events while in altered states of awareness.

Buddhists achieve Nirvana, the awareness of undivided totality or union, through meditation. The ancient tradition of Kabbalah revolves around the contemplation of the 10 Sefirot, or the attributes that God created so he could project himself onto the universe and humankind, in order to achieve a union with Divine Thought, the Ein Soph. Sounding like a description of the Zero Point Field of cutting-edge physics, the Kabbalah states that "Everything shall return to its foundation from which it has proceeded. All marrow, seed and energy are gathered in this place. Hence, all of the potentialities which exist go out through this."

The ultimate union with this field of potentiality is the promise of all major and minor religious traditions, just as it is the promise of scientists seeking the Theory of Everything that would account for the four fundamental forces, and bring together the quantum and gravity. That a human could access this field, and perhaps even achieve oneness with it, is the Holy Grail of spirituality, the brass ring that adepts, sages, and students alike strove to reach out for and grab, through practices of prayer, fasting, drumming, chanting, dancing, meditation, abstinence, music, and other forms of contemplation.

Dr. Wayne Dyer, in his book *The Power of Intention*, suggests that through these avenues one might even be able to affect that field of potentiality, and create their own reality from the power of their thought, or intention. We know from basic physics that the observer's mere presence affects the outcome of experiments with particles and waves. Thought, a form of subtle energy and intention, a more directed "will," may have a resonance or frequency that, like a seed planted in a mound of dirt, is the activating agent of manifestation in the Field. These are metaphysical concepts, to be sure, but they have a basis in the science underlying them. The "field of pure potentiality," as Deepak Chopra calls it, is affected by every thought, memory, intention, action, and experience.

Buddha may have said it better, so I'll let him speak. "All that we are is the result of what we have thought. The mind is everything. What we think, we become." And perhaps what we intend is what we create in our own reality.

Dr. Masaru Emoto's experiments with thoughts and the development of water crystals shows that water responds to our consciousness, even without the spoken word for emphasis. Using signs, Dr. Emoto was able to affect water crystals, forming beautiful crystal shapes in response to positive signs that read "Thank you," and chaotic, uglier crystals in response to a sign stating, "I will kill you." The sound was not necessary. Only the thought or intention of the sign was needed to change the crystals.

Jesus Christ stated, "As you believe, so it will be done to you." This corresponds with the Hermetic teaching of "as above, so below," which unified the macrocosmic with the microcosmic. The explicate with the implicate. Mind with matter. Before Christ walked the Earth 3,000 years ago, Egyptian sage Hermes Trismegistus wrote of the creation of life itself as coming from the Mind of the creator.

"Mind cannot be enclosed, because everything exists within mind. Nothing is so quick and powerful. Just look at your own experience. Imagine yourself in any foreign land and quick as your intention you will be there!" If that isn't referring to remote viewing, I don't know what is!

Hermes goes on to say "All things are thoughts which the Creator thinks." He believed that humans could "see God" by contemplating creation itself. I don't know who coined the phrase "God is in the details," but I am sure Hermes would agree. The presence of the implicate order of David Bohm, the inner workings of the universe, was where we humans could see the handiwork of the creator, as well as the very laws used to create. And similar to the Zero Point Field that Hal Puthoff and other physicists propose is the fundamental energy of the universe, Hermes stated, "The Cosmos is all Life. From its foundations there has never existed a single thing which was not alive." Vibrating, even at close to zero degrees. Space is not a vacuum, but teems with life as subtle, vibratory energy.

Seeing the Light

The most profound spiritual experience of my life also turned out to be the most profound scientifically. I was living in Los Angeles and couldn't find a job. Deeply depressed, I decided to meditate on my problem. Earlier that day, I had spoken to my scientist dad about some book I had read, and we talked about how everything was, at its most fundamental level, made of light. I guess that image stuck in my head, because during my meditation, which was going on much longer than my usual 20-minute stint, I opened my eyes while in a deep state, and everything in the room around me was shapeless and formless waves of vibrating light!

For approximately 10 seconds I stared at my furniture, bed, windows, or at least where they should have been, and instead saw varying frequencies of light, similar to billions of tiny fluctuating particles, yet also waves. Eventually, the light began to fade and solid objects returned in its place, but for the next three days I not only felt physically high as a kite, but also continued to "see" things as being brighter than they usually appeared.

So profound was this experience, I drove up to the local Church of Religious Science on the corner and signed up for classes, determined to become a minister. I told no one at the church about my ambitions and was shocked when someone came up to me as I sat filling out forms and said, "You are going to make a wonderful minister one day." I also intensified my studies into both metaphysics and quantum physics, finding similarities that thrilled me and paralleled my own beliefs and experiences.

Did I see the superposition of everything in my room before the collapse of the wave function? What I thought was just an intense spiritual vision now in retrospect seems like a scientific vision of one of the most basic concepts of quantum physics! Or perhaps I saw the Zero Point Field in its fundamental form, underlying all matter and mass in my bedroom. Whatever happened to me that day *changed my life*, and perfectly mirrors some of the things I later learned about the quantum world.

What Is Out There?

"In my father's house are many mansions," Christ told his disciples. Could this be a reference to parallel universes and extra spatial dimensions? In the Gospel of Thomas, one of the Gnostic texts found in 1945 at Nag Hammadi, Egypt, Jesus also said that the Kingdom of Heaven was spread out upon the earth, but that men do not see it. "It is inside you and outside you . . ." Perhaps this Kingdom resides in an alternate dimension or the ZPF, right at the tip of our very nose, yet we do not have the awareness, or consciousness, to see it.

The Tibetan *Book of the Dead* tells us that we have within us the realms of the higher divisions of the "bardo," and the Indian Vedic tradition believes "The seat of the Gods is indeed within." Laotse said, "For the wise man looks into space and he knows there is no limited dimensions."

Native Americans believe there is no separate and discrete matter that can be measured, that in the quantum world everything is in dynamic flux. There is, in the world of "Blackfoot physics," both manifest and unmanifest reality; the Hopi believe that behind and within all form and matter in nature is the "heart of the Cosmos itself." The unified worldview of the native traditions, both Western and tribal, such as that of Shamans, is the opposite of the typical view of Western science, which separates between matter and mind, outer and inner, above and below.

Shamans could be using parallel universes or the ZPF to journey between the lower, middle, and upper "worlds" to heal the sick and gather information they could not have known with their five senses, their consciousness as the key to shifting from one reality to the other. But quantum physics is coming into full alignment with the spiritual wisdom teachings of ancient, and more modern traditions. The cutting-edge world of the very, very small is opening up a new level of understanding of the greater cosmos, and our place in it, mirroring the beliefs of those who have come before us.

Seeing the Future

As with those who have come before us, those who knew the importance of the implicate order of things, we can use the same techniques as those wise ones and spiritual masters to connect with the part of our consciousness that directly taps into the Cosmic Mind, the Field, the Source, the ether or primordial waters. We may, as a result, achieve just the right resonance to create a paranormal experience, or perceive one that we might otherwise not be aware of. And our minds may be the key. "Mind is the divine part of a human being, which is capable of rising to heaven," says Hermes. Physicist F. David Peat phrased it as such. "Indeed the deeper we dig into the mind the more we discover that the distinction between mind and matter is dissolved and the operation of the objective intelligence begins to manifest its power."

Psychic Jeffrey A. Wands suggests that those desiring a psychic experience develop a mind-set that encourages one. This includes learning to meditate: practice the art of visualization, calm the mind, and most of all, become "open" to having an experience. An agitated or closed mind cannot connect to a higher state of awareness. "The psychic mind-set asks you to look at the world in a completely different way, and that's not always easy. No longer can you take things at face value. No longer can you accept what appears to be. You have to look deeper."

Those who do look deeper, accepting and sharpening their intuition, precognitive skills, and psi abilities will no doubt display even more paranormal abilities, having opened themselves to this new way of "perceiving," that Wands writes about. Believing is seeing, in this case, just as Matthew wrote in the New Testament, "Seek and you shall find, knock and the door shall be opened to you."

I recall the evening before the Northridge Earthquake that hit Los Angeles on January 17, 1994. My husband and I were living in Burbank, just minutes from the city of Northridge. It was late afternoon on the 16th as we drove to a nearby Pavilion's grocery store, and both of us commented on the way the sun just didn't seem right. There was a palpable feeling in the air that something was "wrong," and we verbally expressed our perceptions, joking about the possibility of a big quake coming in the days ahead.

When we got home, our two cats were doing their "earthquake dance," both crouching low to the ground and emitting low howls, and we felt even more certain we had picked up on some quake vibes. We went to bed feeling tentative, and I stayed up for hours reading an Anne Rice novel until I finally fell asleep at 2 AM. A little over two hours later, we were awakened with a jolt and, clinging to each other as the walls shifted and shuddered and groaned around us, wondered if we were going to live through the violent quake that would take the lives of more than 60 people.

What did we feel the night before? Perhaps we had tapped into the Field for just those few moments, where we could "see" the future before it happened in the form of a shift or change in the frequencies our brains were picking up. Animals react to sonic vibrations and frequencies we cannot sense, but maybe, because of our state of minds, we did sense them. Premonitions became reality.

Having a paranormal experience may be a matter of timing and luck, but it may also be a matter of sensitivity to other levels of awareness we don't usually have access to in our busy, crazy lives. A sensory-oriented, intuitive person may be more adept at perceiving things than others, who live more "left brain" lives. Therefore, they may experience more paranormal events such as déjà vu, synchronicities, precognitive dreams, premonitions, hunches, psychic abilities, and ESP. They are, quite simply, open to them.

I meditated for years before I had my son and often had experiences similar to Yogananda's, of a sensation of complete oneness with everything. I referred to them as being "split open" and being everywhere at once, as if my cells just dissipated and spread across the universe. I could literally "see" every corner of space and time. These experiences lasted for less than a minute, but were real and always served to elevate my ability to attract synchronistic events into my life, leading to successes with my career, or to some area of personal growth. In fact, nothing ever clicks in my life until I first get back into that state of being split open. When I am stuck in a sense of my separation from everything else, life tends to suck.

Krishnamurti said, "You are the world." If that's the case, everything the world contains is within you, and that includes psi abilities. The Hindu Vedas and other yogic texts state that the physical, manifest universe is nothing but God's dream. Thoughts in the mind of the Supreme Consciousness. The Logos of creation.

If UFOs and ghosts and paranormal experiences are a product of our minds, or of the world outside of our minds, or something in between, we all have the same latent ability to perceive them. Michael Talbot, author of *The Holographic Universe* calls them "omnijective" experiences; neither subjective nor objective. This term is similar to the "imaginal," used to describe the mystical experiences of Sufis. He suggests UFOs in particular are a product of both an "ultimate lack of division between the psychological and physical worlds. They are indeed a product of the collective human psyche, but they are also quite real." I suggest this applies to all paranormal experiences, and that is what makes them accessible to anyone. Anyone, that is, with a brain, a consciousness, and an open mind, one who understands that "as above, so below."

Structure and essence. The explicate and the implicate. Science and spirit.

Hermes Trismegistus and David Bohm would have been pals.

What Dreams
May Come

*Believe those who are seeking the truth; doubt those who
find it.*

          —Andre Gide

*The world, after all our science and sciences is still a miracle;
wonderful, inscrutable, magical, and more, to whosoever
will think of it.*

         —Thomas Carlyle

As I write this book, scientists claim to have proved
Inflation Theory as fact (until a new fact comes along!), discovered a 10th planet in our solar system bigger than Pluto
(currently called UB313), turned around a few months later
and cruelly demoted poor Pluto to "dwarf planet" status,
measured the gravitational equivalent of a magnetic field for
the first time in a laboratory, and incorporated quantum
effects into a modified theory of gravity. Custom-designed
nanoparticles are being touted as a possible "cure" for some
cancers, and scientists at the Scripps Research Institute have

successfully converted an RNA enzyme (ribozyme) into a DNA enzyme (deoxyribozyme) through a process of "accelerated in vitro evolution." The rate of new discovery is astounding as we develop technologies capable of seeing into the deepest recesses of space, and the smallest corners of the human body.

Nanotechnology whispers of tiny machines that can self-replicate. Quantum computers hint of the possibility of near-light-speed computation. Stem cell research promises the hope of new life to those devastated by disease.

Consciousness researchers buzz with excitement over experiments that may prove our thoughts can directly affect our environment. LEMUR's Genesis Experiment, now underway, will attempt to use collective thought to create life in a tiny test tube filled with nothing but sterile water. And a Massachusetts observatory has just unveiled a powerful new telescope designed to capture possible extraterrestrial light signals. This new project will search the skies of the entire Northern Hemisphere with a type of camera that can detect a billionth-of-a-second flash of light.

I cannot help but think of the famous line from the movie *Field of Dreams*: "If you build it, they will come."

When I was about seven years old, my geophysicist father made the terrible mistake of casually mentioning over supper that our entire neighborhood was once under water millions of years ago. The next day, I came in from the summer sunshine and asked my mom for a spoon. Hours later, my mom looked out a bedroom window and was shocked to see me standing in a hole at least 4 feet deep, a hole that I had cleverly, in Tom Sawyer fashion, enticed some neighbor kids to help me dig. (Well, they did all the digging. I supervised. I was no dummy.)

In addition to finding the main water line that ran under our swing set, I also found dozens of fossils of seashells, which I proudly put into a big glass jar and presented to my father that night when he got home from work.

Two lessons were learned that night. My father learned to watch what he said at the dinner table, and I learned that no theory holds weight until you get your hands dirty, do the work, and put all the pretty rocks in a jar.

I also learned that if you tell me it's there, I will dig for it. If you build it, I will come, spoon and jar in hand. I equate that experience with writing this book. Digging into the areas of the paranormal, of quantum physics, and of new science, I found many rocks that may indeed prove many things. Especially when you put them all in the same pretty jar.

I consider myself a healthy skeptic. My philosophy is "I believe you, now prove it to me." In the case of the fossils, I proved it to myself. I believe the paranormal exists. I have had my own experiences, and met hundreds of people who have had theirs. I also believe that science can explain these mysterious phenomena, although perhaps not just yet. The theories I have presented here certainly hold promise, and some really sound like they can, indeed, explain the existence of other forms of energy present in our reality, energy that uses both actual physical means, and consciousness itself, to get here from there. But similar to many scientists, I seek undeniable proof. It's human nature.

Sometime this decade, physicists will gather at the Large Hadron Collider (LHC) at the particle-physics lab at CERN in Switzerland. Using high-speed collisions of particles, they may prove String Theory is true by finally creating a real string. They may prove M-Theory is correct by detecting evidence of a brane. They may record a trace of extra dimensions. They may prove the existence of "gravitons." Or create antimatter. Or even a tiny black hole.

Or they may find nothing very special at all.

There was a time in our not too distant past when people believed that the Earth was the center of the nearby universe, and that everything else revolved around it. This geocentric model of the universe was shot down in flames when Polish astronomer Nicolaus Copernicus proved that the sun was the center of our solar system, and that the Earth was just one of many planets orbiting it.

The Copernican Principle, as it came to be known, showed us two things. One, that we were nothing special. Two, that knowledge is never absolute. Somebody always comes along to shoot down the other guy's idea of truth. Beverly Rubik, writing in the *British Homeopathic Journal*'s July 1994 edition, put it most succinctly, stating, "The history of science, medicine and technology is full of rejections of novel discoveries that seemed anomalous in their time." She reminds us that Benjamin Franklin was laughed at by his scientist contemporaries when he proposed that lightning was a form of electricity, William Crookes was bitterly attacked by his colleagues when he discovered the element thalium, and that

Lord Kelvin believed that X-rays were nothing but a hoax. "Studies on the psychology of science suggest that scientists have a resistance to acknowledging data that contradict their own hypothesis," she concluded.

Lest you think I am unfairly attacking scientists (whom I admire greatly), realize that religion has its share of extremist thinkers and cultlike followers, rigid and unaccepting of anything that does not fit into their narrow-minded model of reality. The same goes for the paranormal research community, struggling with its own brand of "experts" certain that their theories alone are the absolute, ultimate explanation for a variety of phenomena. Ditto for the medical community, politicians, and just about every other group of people pursuing a common goal you can think of.

George Bernard Shaw once defined opinion as something that is first ridiculed, then blasphemed, then accepted for discussion, then established as a truth. Theory, like opinion, changes and goes through these same stages, but truth remains truth.

It is that truth that I believe we are all seeking.

Science may never fully explain reality. Nor, for that matter, will religion, because reality is ever changing and evolving, morphing into something grander and more utterly and breathlessly complex. Each theory is overthrown by a new assumption, ever more inclusive of all human experience. Maybe we can only hope for a TOE or a GUT that explains our present, but not our future. Maybe we haven't yet envisioned, imagined, and created the reality of our future.

I am now going to commit an act of sheer creative blasphemy and try to improve upon a legend. Remember that quote by William Goldman? "Nobody knows anything." I believe it is a theory that needs to be overthrown, and I am bold enough to do it.

We all know *something*. Paranormal experts know something. Physicists know something. Those working at the cutting edge of brain and consciousness research know something.

Science knows something. So, too, does religion. But nobody knows everything!

Until scientists stop denying any experience that can't be reproduced in a lab or held up against a scientific standard that fails to account for anything yet to be discovered, and until paranormal investigators (AND the religious crowd) stop denying the science behind the "spirit," we continue to get nowhere. Forever stuck in a war of theories, ideas, and assumptions, we only succeed at separating the "normal" from the "paranormal," not understanding that on the deepest level of existence, it's all normal. Rupert Sheldrake, in an interview with Matthew Cromer in 2005, stated his belief that the evidence for paranormal phenomena was greater than the evidence against it, but that scientists needed to open their minds and get out of skeptical denial. "Some 'skeptics' are afraid that if telepathy is admitted to exist, then science as we know it will crumble. This is not the case." In fact, we who are intrigued by the paranormal look to science as the means by which we will one day consider the "unknown" fully known.

Perhaps researchers need to stop asking only questions to which they are likely to have answers. We need to step out of the comfort zone and ask bigger questions, even if it means we may not readily have the answers anytime soon. Even if it means the answers come from places we never thought possible, or acceptable. In my interview with About.com's Paranormal expert Stephen Wagner, he summarized, "To deny the existence of paranormal phenomena just because it cannot be tested by the scientific method is foolish. In a larger sense, such denial is unscientific!"

Perhaps these mavericks of quantum physics, consciousness research, and the paranormal that I write about in this book will one day bring us to the knowledge, and the truth, about our true nature of connectedness to one another, and to the universe we live in, and remind us who we are, what we are made of, why we are here, and where we are going.

I hope that this book can be a small part of that ongoing dialogue in a world that cries out for unity, harmony, understanding and, most of all, resonance. Oh, and truth.

Because after all the arguing and theorizing and debating, we discover that we really are all on the same quest, heading in the same direction . . . to find the same Holy Grail.

Happy hunting!

antigravity Anything that cancels the effect of gravity.

antimatter An undetected entity that cancels out the properties of matter.

Big Bang A term used to describe the explosion that marked the birth of the universe.

black hole A collapsed star with such density that nothing with mass can escape its gravity.

curved space The theory that the contours of space follow those of the gravitational fields that occupy it.

fourth dimension Time.

gravity A force that originates from and attracts all objects with mass.

paranormal Something that is beyond the scope of being normal.

physics The science that studies matter and energy.

relativity The measuring of the position or speed of one object in relation to that of another.

String Theory A theory in quantum mechanics that postulates that the universe is made up of bands, or frequencies, of energy nicknamed strings.

theory A scientific postulation that has not been proven.

worm hole A theoretical "tunnel" in the space/time continuum through which, it is believed, parallel universes can be accessed.

The American Astronomical Society
American Geophysical Union (AGU) Building
2000 Florida Avenue NW, Suite 400
Washington, DC 20009-1231
(202) 328-2010
Web site: http://www.aas.org
Established in 1899, the American Astronomical Society
    (AAS) is an organization of professional astronomers in
    North America.

American Institute of Physics
One Physics Ellipse
College Park, MD 20740-3843
(301) 209-3100
Web site: http://www.aip.org
The American Institute of Physics is dedicated to the
    advancement of physics and dissemination of information
    pertaining to the science.

American Physical Society
American Center for Physics
One Physics Ellipse

College Park, MD 20740

Web site: http://www.aps.org

A society of scientists dedicated to the advancement and diffusion of
the knowledge of physics.

International Astronomical Union

IAU - UAI Secretariat

98-bis Blvd Arago

F–75014 Paris

France

Web site: http://www.iau.org

An international organization of scientists devoted to the study of
astronomy.

NASA (National Aeronautics and Space Administration)

Public Communications Office

NASA Headquarters

Suite 5K39

Washington, DC 20546-0001

(202) 358-0001

Web site: http://www.nasa.gov

At the Web site of the central space organization in the United States
you can find information about current space programs and
research.

The Planetary Society

65 North Catalina Avenue

Pasadena, CA 91106-2301

(626) 793-5100

Web site: http://www.planetary.org

The Planetary Society involves itself with the study and dissemination
of information on the latest planetary news.

UFO Magazine

P.O. Box 11013

Marina del Rey, CA 90295

Web site: http://www.ufomag.com
UFO is an online publication that tracks paranormal phenomena in
   the realm of UFOs and unknown spacecraft.

Web Sites

Due to the changing nature of Internet links, Rosen Publishing has
developed an online list of Web sites related to the subject of this
book. This site is updated regularly. Please use this link to access
the list:

http://www.rosenlinks.com/hgp/msp

Greene, Brian. *The Elegant Universe: Superstrings, Hidden Dimensions, and the Quest for the Ultimate Theory.* New York, NY: W. W. Norton & Company, 2003.

Hawes, Jason. *Ghost Hunting: True Stories of Unexplained Phenomena from The Atlantic Paranormal Society.* New York, NY: Pocket, 2007.

Kaku, Michio. *Parallel Worlds: A Journey Through Creation, Higher Dimensions, and the Future of the Cosmos.* New York, NY: Anchor Books, 2006.

Radin, Dean. *Entangled Minds: Extrasensory Experiences in a Quantum Reality.* New York, NY: Paraview Pocket Books, 2006.

Randall, Lisa. *Warped Passages: Unraveling the Mysteries of the Universe's Hidden Dimensions.* New York, NY: Ecco, 2005.

Van Praagh, James. *Ghosts Among Us: Uncovering the Truth About the Other Side.* New York, NY: HarperOne, 2008.

Al-Khalili, Jim. *Black Holes, Wormholes and Time Machines.* London, England: Institute of Physics Publishing, 1999.

Arntz, William, Betsy Chase, and Mark Vincente. *What the BLEEP Do We Know!?* Deerfield Beach, FL: Health Communications, Inc., 2005.

Astronomy, "Time Travel." February 2006.

Belanger, Jeff. *The World's Most Haunted Places: From the Secret Files of Ghostvillage.com.* Franklin Lakes, NJ: New Page Books, 2004.

Bem, Darryl J., and Charles Honorton. "Does Psi Exist?" *Psychological Bulletin*, Vol. 115, 1994.

Bord, Colin and Janet Bord. *Unexplained Mysteries of the 20th Century.* Chicago, IL: Contemporary Books, 1989.

Calder, Nigel. *Einstein's Universe: The Layperson's Guide.* London, England: Penguin Books, 2005.

Clark, Jerome. *Unexplained!* Detroit, MI: Visible Ink Press, 1993.

Cochrane, Hugh F. "The Great Lakes Triangle—The Mystery Continues." *UFO Report*, August 1980.

Coleman, Jerry D. *Strange Highways: A Guidebook to American Mysteries & The Unexplained.* Decatur, IL: Whitechapel Productions Press, 2003.

Cook, Nick. *The Hunt for Zero Point: Inside the Classified World of Antigravity Technology*. New York, NY: Broadway Books, 2001.

Dowling, Jason. "Bermuda Triangle Mystery Solved? It's a Load of Gas." *Fairfax Digital*, October 2003.

Dyer, Dr. Wayne W. *The Power of Intention*. Carlsbad, CA: Hay House, 2004.

Emoto, Masaru. *The Secret Life of Water*. New York, NY: Atria Books, 2005.

Fawcett, Lawrence, and Barry Greenwood. *Clear Intent: The Government Cover-Up of the UFO Experience*. Englewood Cliffs, NJ: Prentice-Hall, Inc., 1984.

Gooch, Stan. *The Dream Culture of the Neanderthals: Guardians of the Ancient Wisdom*. Rochester, VT: Inner Traditions, 2006.

Good, Timothy. *Above Top Secret: The Worldwide UFO Cover-up*. New York, NY: William Morrow, 1988.

Goswami, Amit. *Self-Aware Universe: How Consciousness Creates the Material World*. New York, NY: Tarcher, 1995.

Greene, Brian. *The Elegant Universe: Superstrings, Hidden Dimensions and the Quest for the Ultimate Theory*. New York, NY: Vintage, 2000.

Greene, Brian. *The Fabric of the Cosmos*. New York, NY: Knopf, 2004.

Holzer, Hans. *The Supernatural: Explaining the Unexplained*. Franklin Lakes, NJ: New Page Books, 2003.

Kaku, Michio. *Hyperspace: A Scientific Odyssey Through Parallel Universes, Time Warps and the 10th Dimension*. New York, NY: Oxford University Press, 1994.

Kaku, Michio. *Parallel Worlds: A Journey Through Creation, Higher Dimensions and the Future of the Cosmos*. New York, NY: Doubleday, 2005.

Kenyon, J. Douglas, et al. *Forbidden History: Prehistoric Technologies, Extraterrestrial Intervention and the Suppressed Origins of Civilization*. Rochester, VT: Bear and Co., 2005.

La Violette, Paul A. *Genesis of the Cosmos: The Ancient Science of Continuous Creation*. Rochester, VT: Bear & Co., 2004

Livio, Mario. *The Accelerating Universe: Infinite Expansion, the Cosmological Constant and the Beauty of the Cosmos*. New York, NY: John Wiley and Sons, 2000.

Lloyd, Seth. *Programming the Universe: A Quantum Computer Scientist Takes on the Cosmos*. New York, NY: Knopf, 2006.

Marrs, Jim. *Alien Agenda: Investigating the Extraterrestrial Presence Among Us*. New York, NY: HarperTorch, 1998.

Marrs, Jim. *PSI Spies*. Phoenix, AZ: AlienZoo, Inc., 2000.

McTaggart, Lynne. *The Field: The Quest for the Secret Force of the Universe*. New York, NY: HarperCollins, 2002.

Novak, Peter. *The Lost Secret of Death: Our Divided Souls and the Afterlife*. Charlottesville, VA: Hampton Roads, 2003.

Ouellette, Jennifer. *Black Bodies and Quantum Cats: Tales from the Annals of Physics*. New York, NY: Penguin Group, 2005.

Pagels, Heinz R. *The Cosmic Code: Quantum Physics as the Language of Nature*. New York, NY: Simon & Schuster, 1982.

Panek, Richard. *The Invisible Century: Einstein, Freud and the Search for Hidden Universes*. New York, NY: Penguin Books, 2004.

Parodi, Angelo. *Science and Spirit: What Physics Reveals About Mystical Belief*. Uniondale, PA: Pleasant Mount Press, 2005.

Peat, F. David. *Synchronicity: The Bridge Between Matter and Mind*. New York, NY: Bantam, 1987.

Penrose, Roger. *The Emperor's New Mind*. New York, NY: Oxford University Press, 1989.

Quasar, Gian J. *Into the Bermuda Triangle: Pursuing the Truth Behind the World's Greatest Mystery*. New York, NY: International Marine/McGraw Hill, 2004.

Radin, Dean. *Entangled Minds: Extrasensory Experiences in a Quantum Reality*. New York, NY: Paraview Pocket Books, 2006.

Randles, Jenny. *Breaking the Time Barrier: The Race to Build the First Time Machine*. New York, NY: Paraview Pocket Books, 2005.

Randles, Jenny. *Time Storms*. New York, NY: Berkley, 2001.

Redfern, Nick. *Three Men Seeking Monsters: Six Weeks in Pursuit of Werewolves, Lake Monsters, Giant Cats, Ghostly Devil Dogs and Ape-Men*. New York, NY: Paraview Pocket Books, 2004.

Redfern, Nick, and Andy Roberts. *Strange Secrets: Real Government Files on the Unknown*. New York, NY: Paraview Pocket Books, 2003.

Rees, Martin. *Our Cosmic Habitat*. Princeton, NJ: Princeton University Press, 2001.

Roach, Mary. *Spook: Science Tackles the Afterlife*. New York, NY: W. W. Norton, 2005.

Roll, William, and Valerie Story. *Unleashed: Of Poltergeists and Murder: The Curious Story of Tina Resch*. New York, NY: Paraview Pocket Books, 2004.

Russell, Peter. *From Science to God: The Mystery of Consciousness and the Meaning of Light*. Sausalito, CA: Peter Russell, 2000.

Scientific American: Special Edition, "The Frontiers of Physics," February 2006.

Seed: Science Is Culture Magazine, February/March 2006.

Seife, Charles. *Decoding the Universe: How the New Science of Information Is Explaining Everything in the Cosmos, from Our Brains to Black Holes*. New York, NY: Viking, 2006.

Sever, Megan. "Beneath the Bermuda Triangle." *GeoTimes*, November 2004.

Sheldrake, Rupert, Terence Mckenna, and Ralph Abraham. *The Evolutionary Mind: Conversations on Science, Imagination and Spirit*. Rhinebeck, NY: Monkfish Books, 2005.

Siegfried, Tom. *The Bit and the Pendulum: From Quantum Physics to M-Theory—The New Physics of Information*. New York, NY: John Wiley, 2000.

Siegfried, Tom. *Strange Matters: Undiscovered Ideas at the Frontiers of Space and Time*. Washington, DC: Joseph Henry Press, 2002.

Sullivan, Randall. "Spirited Away." *Reader's Digest*, February 2006.

Swanson, Claude, Ph.D. *The Synchronized Universe: New Science of the Paranormal*. Tuscon, AZ: Poseidia Press, 2003.

Talbot, Michael. *The Holographic Universe*. New York, NY: HarperPerennial, 1991.

Talcott, Richard. "Is Time on Our Side?" *Astronomy*, February 2006.

Vallee, Jacques. *Dimensions: A Casebook of Alien Contact*. Chicago, IL: Contemporary Books, 1988.

Wands, Jeffrey A. *The Psychic in You: Understand and Harness Your Natural Psychic Powers*. New York, NY: Pocket Books, 2004.

Warren, Joshua P. *How to Hunt Ghosts: A Practical Guide*. New York, NY: Fireside Books, 2003.

Warren, Joshua P. *Pet Ghosts: Animal Encounters from Beyond the Grave*. Franklin Lakes, NJ: New Page Books, 2006.

INDEX

A

Abbott, Edwin, 137
abductions, alien, 32
Abraham, Ralph, 155
Air Force Museum, 36
Akashic Records, 36, 87, 156, 178, 188
Albrecht, Andy, 125
alien contact, 20
allotropic forms of hydrogen, 102
American Antigravity, 151–153
Andreasson, Betty, 33
antigravity, 168, 172–174
antimatter, 172–174
Arguelles, Jose, 214
Arnold, Kenneth, 28
Asimov, Isaac, 135
Aspect, Alain, 124, 202
astral travel, 80
astronauts, UFO sightings by, 32
Avro Canada, 149
Avrocar, 36–38, 149

B

Babylonians, 89, 218
Ball, Phillip, 20
Banks, Joseph, 81
Batts, Kate, 49

Belgium, UFO sighting in, 31
Bell Witch, 49
Bell, John S., 49, 109
   Theorem, 109, 156
Benioff, Paul, 133
Bermuda Triangle, the, 63–73, 74, 77, 168, 175–176
Big Bang, 114, 125, 219
Big Bounce theory, 122
Big Splat, the, 114
Bigfoot, 58, 60, 74
Black Angus, 61
Black Hole thermodynamics, 128
black holes, 132, 160–161
Black Mausoleum, 48
Black Shuck, 61
Blackfoot physics, 224
Blanke, Olaf, 86
b-mesons, 101
Bohm, David, 127, 129, 189, 214, 227
Bohr, Niels, 102, 103, 108
Bondi, Hermann, 173
Book of Revelations, the, 89
Booth, John Wilkes, 87
Bord, Janet and Collin, 57
Born, Max, 102, 103
Bose, Satyendra Nath, 100
Bose-Einstein Condensates, 104
bosons, 101

ABOUT THE AUTHOR

Marie D. Jones is a lifelong student, researcher, and investigator of metaphysics and the paranormal. She is the author of *Looking for God in All the Wrong Places*, as well as hundreds of published articles, book reviews, essays, and short stories. She lives in San Marcos, CA.